# The Blood–Brain Barrier, Amino Acids and Peptides

# The Blood–Brain Barrier, Amino Acids and Peptides

by

**M.B. Segal PhD MPS**
*Reader in Physiology*
*Sherrington School of Physiology*
*United Medical and Dental Schools of Guys and*
*St Thomas' Hospital*
*London*
*England*

and

**B.V. Zlokovic MD PhD**
*Associate Professor of Medical Physiology*
*Faculty of Medicine*
*Belgrade*
*Yugoslavia*

**Present address:**
*Research Associate Professor of Neurosurgery*
*Childrens Hospital of Los Angeles*
*University of Southern California*
*Los Angeles*
*USA*

KLUWER ACADEMIC PUBLISHERS
DORDRECHT / BOSTON / LONDON

**Distributors**

*for the United States and Canada:* Kluwer Academic Publishers, PO Box 358, Accord Station, Hingham, MA 02018-0358, USA
*for all other countries:* Kluwer Academic Publishers Group, Distribution Centre, PO Box 322, 3300 AH Dordrecht, The Netherlands

**British Library Cataloguing in Publication Data**

Segal, M.B.
The Blood–brain barrier, amino acids and peptides.
1. Mammals. Blood–brain barrier. Amino acids. Peptides
I. Title II. Zlokovic, B.H.
599.01′88

ISBN-13: 978-94-010-7500-8 e-ISBN-13: 978-94-009-2229-7
DOI: 10.1007/978-94-009-2229-7



Published in the United Kingdom by Kluwer Academic Publishers, PO Box 55, Lancaster, UK.

Kluwer Academic Publishers BV incorporates the publishing programmes of D. Reidel, Martinus Nijhoff, Dr W. Junk and MTP Press.

Lasertypeset by Martin Lister Publishing Services, Carnforth, Lancs.

# Contents


| | | |
|---|---|---|
| 1 | Introduction | 1 |
| 2 | Methods for the study of the functions of the blood–brain barrier | 13 |
| 3 | Amino acids and the blood–brain barrier | 31 |
| 4 | The role of peptides in the brain | 47 |
| 5 | Brain peptides and peptidergic pathways | 91 |
| 6 | The interaction of small peptides with the various cellular interfaces of the blood–brain barrier | 123 |
| 7 | Transport of large peptides and proteins across the blood–brain barrier | 149 |
| 8 | Disease states and the blood–brain barrier related to peptides | 165 |
| 9 | Immunological and other aspects of the blood–brain barrier | 179 |
| 10 | Strategies for peptide delivery to the brain | 189 |
| | Index | 193 |

# Acknowledgement

The suggestion for this review came from my dear friend, Professor Mortyn Jones, the neuroendocrinologist, who tragically died of cancer at the peak of his career. I would like to dedicate this review to him and I am sorry he is not with us to give us the benefit of his razor sharp mind. Since my main interest is amino acids, I have asked my colleague Berislav Zlokovic to join me in writing this review, which he has done with great skill, leaving me with less to write and more to edit than I expected.

We both wish to thank our great friends and mentors, Academician Professor Hugh Davson and Ljubisa Rakic, without whom our progress in science would have been limited. We also wish to thank Ms. Fiona Donaldson who typed the manuscript and tolerated the many changes we made.

*Malcolm Segal*
*Berislav Zlokovic*

# 1 Introduction

## Definition of the barrier

The interstitial fluid (ISF) of the brain is separated from the blood by the blood–brain barrier (BBB). This barrier must not be thought of as a single entity or as an absolute restriction to all molecules, but as a multiple structure located at several sites within the brain. The first of these interfaces is located at the endothelium of the brain capillaries. Secondly there is a potential site for interchange on the outer linings of the brain between the dura and the arachnoid membranes. Thirdly, there are the choroid plexuses and the cerebrospinal fluid (CSF) which is in contact with the very permeable internal ependymal lining of the brain (Figure 1.1). Finally there are areas of the brain which lack a blood–brain barrier. These areas constitute the circumventricular organs and have leaky capillaries with a barrier at the ependyma which limits the spread of molecules from those regions (Figure 1.2). Each of these BBB sites has its own characteristic permeability and transport functions. We will now consider the properties of each of these barrier sites between the blood and the brain.

## THE CEREBRAL CAPILLARY ENDOTHELIUM

### Morphology

This interface has both the largest surface area and the shortest path length between the blood and the brain ISF. The cerebral capillaries, which at first sight seem little different from the rest of the systemic circulation, are in fact unique to the brain. Systemic capillaries are divided into two types (a) fenestrated, and (b) continuous, which refer to their observed morphological appearance.

The majority of the cerebral circulation is composed of continuous capillaries where the endothelium forms a complete double layer and contains a reasonable quantity of cytoplasm and intracellular organelles. These capillaries do, however, differ from other continuous capillaries in that the cytoplasm of the cerebral endothelial cells contains many more mitochondria

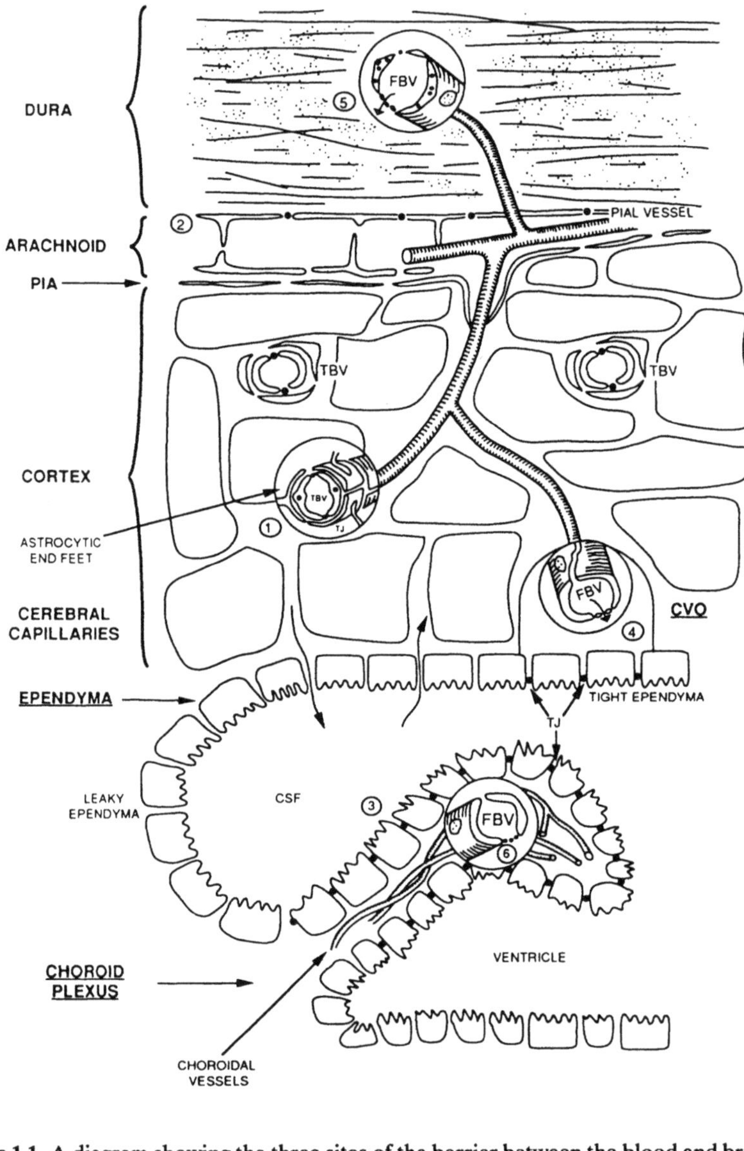


**Figure 1.1** A diagram showing the three sites of the barrier between the blood and brain. (1) The cerebral capillaries (TBV) with tight junctions (TJ) sealing the endothelial cell margins. (2) The outer layer of the arachnoid also with TJ. (3) The epithelium of the choroid plexus joined by TJ except at the base. (4) Certain regions lack a barrier, the circumventricular organs (CVO) with fenestrated capillaries (FBV). For ependyma is leaky except in regions close to the CVOs where there are TJs. (5) Dural vessels are outside the barrier. (6) FBV are also found in the choroid plexus

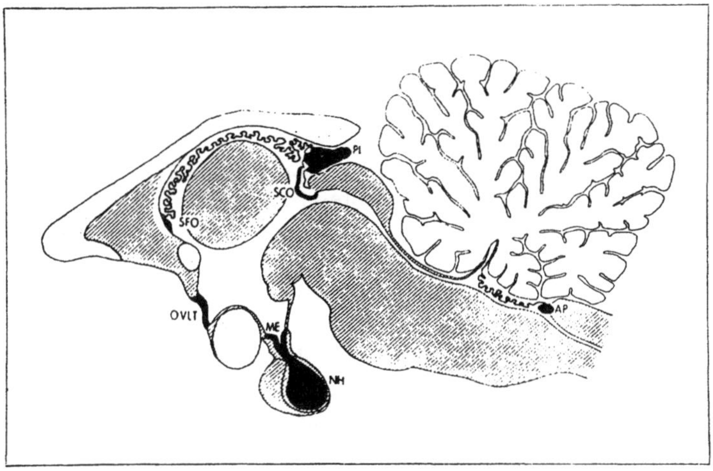


**Figure 1.2** Section through squirrel monkey brain in median sagital plane. Most parts of venticular walls are covered by ciliated ependyma. Specialized non-ciliated ependymal cells cover circumventricular organs. These and similarly non-ciliated epithelium of choroid plexus of third and fourth ventricle are outlined in black. Located around third venticle are following circumventricular organs: median eminence (ME), neurohypophysis (NH), pineal body (PI), organum vasculosum of the lamina terminalis (OVLT), subfornical organ (SFO), and subcommissural organ (SCO). Area postrema (AP) is found at transition of fourth ventricle into central canal. (From Weindl and Joynt, 1972[55] with permission)

than any other capillaries, which may indicate that these cells are capable of considerable metabolic work[1]. Another feature is that cerebral capillaries often appear to be running through the middle of the endothelial cell, so these vessels are formed from a continuous 'tube' of endothelium with no 'seams'[2]. Also, in other systemic capillaries there is a gap between endothelial cells which constitutes a low resistance pathway for all small molecules between blood and ISF. In contrast, in the brain these 'gaps' between the cells of capillary endothelium are obstructed by a continuous belt of occluding tight junctions which are sites of high transendothelial resistance[3,4]. Finally, the most unusual feature of cerebral capillaries is that they are virtually enveloped by a 'tube' composed of astrocytic end feet which are projections of specialized glia. Many of these glia also have processes in contact with the ependymal lining of the ventricles and so appear to be a connection between the blood, the cerebrospinal fluid (CSF) and the brain ISF. The function of these glia is still a matter of debate but at present no firm conclusions can be drawn. However, it is now known that the astrocytic end feet do not form a continuous layer around the cerebral capillaries and electron dense macromolecules in the CSF can pass into the subcapillary space, so they do not constitute an additional barrier to the diffusion of substances from the blood to the brain ISF[5]. The cerebral circulation only has fenestrated capillaries in the choroid plexuses and circumventricular organs, which are regions with special functions and will be discussed later. The dural capillaries are also fenestrated, but these lie outside the blood–brain barrier[6]. Fenestrated capillaries have regions of endothelial wall where the cytoplasm is excluded and the two sides of the endothelial wall have become fused. In these areas there are circular patches, the fenestrae, where the wall is extremely thin or absent. These fenestrae have been proposed as leak pathways with a high hydraulic conductivity[7] but this is disputed by some[8,9].

## Permeability characteristics of cerebral capillaries

Krogh, in his 1948 Croonian address[10], likened the permeability characteristics of the blood–brain barrier to those of the cell wall. Substances which have a high oil/water partition coefficient, for example anaesthetics, fat soluble molecules and respiratory gases, cross easily and rapidly gain access to the brain from the blood[11,12]. In contrast, polar molecules exhibit various degrees of restricted permeability depending on their properties and the selectivity offered by the capillary endothelium. The simplest property of a molecule is its size and shape and in the case of the BBB even quite small molecules, such as glycerol and urea, have a slow rate of permeation into the brain. Urea, for example, penetrates so slowly that it has been used as a hypertonic intravenous fluid to withdraw water from the brain if cerebral oedema occurs during neurosurgery. Considering the fact that urea will equilibrate fully across a red cell membrane, this illustrates the remarkable restriction offered by the BBB[13,14]. Molecules such as sucrose (with a molecular weight of 300) are virtually excluded by this barrier[15]. In contrast, molecules such as D-glucose and various L-amino acids easily gain access to the brain, whereas others such as mannitol, L-glucose and glycine do not[16]. Since these permeating molecules demonstrate saturation, stereospecificity and competition, there can be little doubt that this permeation is a carrier mediated process[17]. Electrolytes are similarly obstructed by the BBB, which further demonstrates that size and charge play a crucial part in the restrictive nature of the BBB. Sodium, for example, will equilibrate with muscle within minutes yet takes 24–48 hours to fully equilibrate with the brain[18].

## Routes of permeation through the cerebral endothelium

There are three possible routes through the wall of the brain capillary. Firstly molecules may pass directly through the wall and, as we discussed in the last section, this route depends primarily on the nature of the cell wall and its selectivity. Secondly, molecules may pass through intercellular clefts between endothelial cells. Studies with electron dense molecules, such as horseradish peroxidase, colloidal lanthanum, and the smallest of these molecules, microperoxidase (1800 Daltons), show that they all fail to penetrate the intercellular clefts between endothelial cells of the cerebral capillary wall[3,4]. These studies have shown that the point of obstruction is the occluding band of tight junctions, the zonula occludens, which joins the endothelial cells together.

Freeze fracture studies, using scanning electron microscopy, have resolved the nature of these junctions into a series of strands; apparent correlation was found between the number of strands within the junctional complex and the resistance offered by the cell sheet[19–21]. However, subsequent studies have challenged this simple concept and tortuous open pathways have been found between the strands of some zonula occludens using serial sectioning techniques[21]. These morphological studies are limited by the molecules which can be visualized under the electron microscope and can only place a lower estimate on the size of this potential water-filled pathway as being less than 1.5 nm. However, since sugars, amino acids and electrolytes are much smaller than this and exhibit a limited permeability, the dimensions of the pathway for free diffusion across the blood–brain barrier must be less than 0.5 nm[22].

As has been stated, the permeability of the cerebral capillaries is many orders smaller than that of other systemic beds. Michel and his co-workers have shown that, although there is a potential open pathway between the endothelial cells of most capillary beds, this route is in fact filled with a fibrillar matrix of basement membrane. This matrix is charged and can restrain macromolecules such as albumin, even though the dimensions of the morphological path appear much greater[23,24]. In the brain the nature of this fibrillar matrix has not been considered and studies have concentrated primarily on the tight junctions. In passing, it is interesting to note that cerebral capillaries offer a considerable resistance to the movement of small molecules, so that the Starling equilibrium cannot apply to the brain. Having capillaries of such high resistance does permit these structures to limit the species of molecule which can cross the cerebral endothelium and it is this obstruction that removes the need for a lymphatic drainage system within the brain.

The third potential pathway is by the vesicles, seen within the cytoplasm and occasionally fusing the endothelial wall. The problem has been to resolve the dynamics of these structures and whether there is a net direction of movement. There can be little doubt that substances on either side of the cell can gain access to vesicles within the cytoplasm, but no one has as yet demonstrated a net direction of movement of these vesicles within the cytoplasm. The difficulty with these morphological studies is related to the slow nature of the fixation process and the volume of sections which need to be cut to serially section large areas of the capillary wall. Some prodigious studies have reconstructed the interaction of the vesicles with the cell wall and have demonstrated a potential path for macromolecules[25,26]. These studies have demonstrated chains of vesicles forming a continuous pathway across the cell, but these findings are vigorously challenged by other workers in the field[27,28]. Recent studies with ultrathin sections following gold colloid may resolve this dispute[29]. However, if it is assumed that vesicles may contribute to macromolecular passage across the blood–brain barrier, it must not be assumed that this route is a non-specific bulk pathway, but one which appears to require that the macromolecule interacts with a specific membrane-bound receptor and the receptor molecule complex is then internalized. Experimental evidence, which will be discussed in Chapter 8, has demonstrated a receptor mediated uptake for some specific macromolecules such as the cationized form of albumin as well as a time dependent uptake of IgG. These findings now need to be correlated with morphological observations to determine the route that these molecules have taken in their passage from blood to the brain ISF.

Future studies must combine physiological, immunological and morphological studies to resolve the mechanism by which macromolecules can cross the physical restriction offered by the BBB and enter the brain. As will be discussed later, the physiologists' view of a capillary wall acting solely as a fixed cellular tube is now no longer tenable and the dynamic nature of endothelia, long familiar to the immunologists, must now be integrated into the functions of the blood–brain barrier.

## Apparent barrier mechanisms – enzymes

A variety of neurotransmitters and peptides have one set of actions in the blood and a completely different role in the brain. For example, intravenous norepinephrine in the chicken causes excitement whereas if injected into the CSF it causes sleep[30]. This type of dual function requires that these agents in the blood be excluded from the brain. At first sight it seemed that this was achieved by the simple physical restriction of the BBB. However, studies with labelled molecules have shown that the label was able to penetrate from blood to brain. More careful studies have revealed that although minute trace amounts do cross the BBB, the bulk of the compound is degraded either by enzymes in the blood or by specific enzymes bound to the wall of the cerebral capillaries[31,32].

This type of study can easily lead to misinterpretation, since when molecules such as peptides are degraded to their constituent amino acids which can be labelled and are easily transported, they will give the appearance of an uptake of the peptide[33]. It is thus always essential to check the integrity of the macromolecule by a specific identification process, before the uptake of these molecules can be substantiated (see Chapter 8).

## THE DURA, ARACHNOID AND PIA

The external coverings of the brain, the meninges, are composed of the dura, the arachnoid and the pia and form a potential interface with the external surface of the brain and the spinal cord. The dura contains fenestrated capillaries so that large molecules can pass from the blood into the dura. However, further passage towards the brain is limited by the external layer of the arachnoid, the cells of which are joined together by tight junction[34]. The blood vessels within the pia are all of the continuous type seen elsewhere in the brain so constituting the BBB in these regions. However, studies on isolated pieces of the arachnoid membrane and with *in vivo* perfusions of the ventricle to the subarachnoid space do point to this layer as having some transport capabilities[35,36]. There is, however, a problem in ensuring that the needles used to perfuse or to collect from the subarachnoid space are actually beneath the outer arachnoid membrane and not in the subdural space. The same problem may be encountered with cups on the brain surface which have been used to study uptake by the brain *in vivo*[37]. If there is an intact outer arachnoid membrane in these methods there will be little movement towards the brain since this layer is the site of the BBB at this interface. However, a large clearance of substances may still be observed, since if the fluid is in contact with the dural vessels, which are outside the BBB, molecules will pass easily into the bloodstream as these vessels are of the permeable systemic type[38].

## CHOROID PLEXUSES AND CSF

The ventricles and subarachnoid space of the brain and spinal cord are filled with cerebrospinal fluid (CSF). This fluid, which is formed by the choroid plexuses within the ventricles, slowly flows through these cavities into the subarachnoid spaces where it drains back into the blood via the arachnoid granulations, the spinal nerve roots and the olfactory tracts[39–41].

The choroid plexuses are well vascularized organs; those in the lateral and third ventricles are double-sided leaf-like structures, whereas that in the fourth ventricle is a single-sided structure dividing this ventricle from the cisterna magna. The capillaries of the choroid plexuses are of the 'leaky' fenestrated type, but although macromolecules can gain access to the interstitial space of the plexuses, further progress towards the CSF is halted by an occluding band of tight junctions joining the cells of the choroidal epithelia together close to the CSF side of the tissue[42–44]. The choroid plexuses are less of a barrier than that found within the capillaries of the brain, so molecules can pass slightly more easily from blood to CSF than from blood to brain ISF[45,46]; however, the area of interface offered by the choroidal interface is 1/5000 of that of the cerebral capillaries. These structures, as well as secreting CSF, can secrete proteins and are capable of transporting electrolytes, sugars, amino acids and other molecules between blood and CSF, so forming an important element of the brain homeostatic mechanism[47–49].

It is of interest that the cells which cover the choroid plexuses are continuous with the ependymal linings of the ventricles. Many studies have shown that the ependyma is an extremely leaky membrane so that substances in the CSF freely exchange with the brain ISF[50].

At the base of the choroid plexuses, there is a sudden transition between 'tight' choroidal epithelia and the leaky ependyma. Since the choroidal blood vessels are fenestrated, there is a potential leak pathway which may account for the entry of lanthanum from blood into CSF by this route[51,52]. However, since the concentration of macromolecules is low in CSF, this route must be limited primarily by the long path length and slow diffusion, so in fact it makes little real contribution to the exchanges between blood and CSF.

## SPECIAL REGIONS OF THE BRAIN. THE CIRCUMVENTRICULAR ORGANS

Whilst the vast majority of the capillaries of the brain are of the continuous 'tight' type, a few sites do exist, where the capillaries are 'leaky' and the barrier absent. These sites constitute a group of special brain regions, the circumventricular organs[53], so called since they are all located close to the ventricles of the brain. These regions were originally identified by dye studies and staining was seen in the area postrema, median eminence, neuro hypophysis, pineal gland, the organum vasculosum of the lamia terminalis and the subcommisural organ as well as the choroid plexuses. These structures are now known to be involved in the salt balance, hormonal feedback and release processes and are under intense investigation (see Chapter 4). It is thus obvious that free exchange is necessary between blood and brain for their functions, but these sites do not constitute a route of free access to other regions of the brain, since the area is bounded by a region of ependyma with tight junctions, so limiting escape of molecules from these organs[54].

## CONCLUSION

The blood–brain barrier is a multiple structure situated at the cerebral capillary endothelium, the choroid plexuses and the outer arachnoid membrane. Each of these sites can act as a simple physical barrier, restraining the movement of molecules and ions, the degree of restraint being largely governed by lipid solubility.

In addition to this simple physical barrier, these interfaces contain a variety of transport proteins, which can move electrolytes, sugars, amino acids and other non-electrolytes between the blood and brain and CSF, often against an existing concentration gradient. Enzymes in blood and those bound to the cerebral endothelium can denature molecules attempting to cross the BBB.

Finally, these structures may also transport intact macromolecules by specific processes, the mechanism of which is at present poorly understood.

## REFERENCES


1. Oldendorf, W.H. and Brown, W.J. (1975). Greater number of capillary and endothelial cell mitochondria in brain than in muscle. *Proc. Soc. Exp. Biol. Med.*, **149**, 736–738
2. Oldendorf, W.H., Cornford, M.E. and Brown, W.J. (1977). The large apparent work capability of the blood brain barrier: a study of the mitochondrial content of capillary endothelial cells in brain and other tissues. *Ann. Neurol.*, **1**, 409–417
3. Reese, T.S. and Karnovsky, M.J. (1967). Fine structural localization of a blood–brain barrier to exogeneous peroxidase. *J. Cell Biol.*, **34**, 207–217
4. Feder, N. (1971). Microperoxidase an ultrastructural tracer of low molecular weight. *J. Cell Biol.*, **51**, 339–343
5. Brightman, M.W. (1967). The intracerebral movement of proteins injected into blood and cerebrospinal fluid of mice. *Prog. Brain Res.*, **29**, 19–37
6. Andres, K.H. (1967). Uber die Fienstanktur der Arachnordea und Duramater von Mammalie. *Z. Zellforsch.*, **79**, 272–295
7. Simionescu, N., Simionescu, M. and Palade, G.E. (1972). Permeability of intestinal capillaries pathway followed by dextrans and glycogens. *J. Cell Biol.*, **53**, 365–392
8. Renkin, E.M. (1978). Transport pathways through capillary endothelium. *Microvasc. Res.*, **15**, 123–135
9. Artuson, G. and Granath, K. (1972). Dextrans as test molecules in studies of functional ultrastructure of biological membranes. *Clin. Chim. Acta*, **37**, 309–322
10. Krogh, A. (1946). The active and passive exchanges of inorganic ions through the surfaces of living cells and through living membrane generally. *Proc. R. Soc. Biol.*, **133**, 140–200
11. Rappoport, S.I., Ohno, K. and Pettigrew, K.O. (1979). Drug entry into the brain. *Brain Res.*, **172**, 354–359
12. Cornford, E.M., Braun, L.D. and Oldendorf, W.H. (1982). Comparison of lipid mediated blood–brain permeability in neonates and adults. *Am. J. Physiol.*, **243**, C161–168
13. Kleeman, C.R., Davson, H. and Levin, E. (1962). Urea transport in the central nervous system. *Am. J. Physiol.*, **203**, 739–747
14. Crone. C. (1965). The permeability of brain capillaries to non-electrolytes. *Acta Physiol. Scand.*, **64**, 407–417
15. Davson, H. and Bradbury, M.W.B. (1965). The extracellular space of the brain. In: *Biology of Neuroglia*, eds., E.D.P. De Robertis and R. Carrea. (Elsevier, Amsterdam)
16. Oldendorf, W.H. (1971). The brain uptake of radiolabeled amino acids, amines and hexoses after intra arterial injection. *Am. J. Physiol.*, **221**, 1629–1639
17. Oldendorf, W.H. and Szabo, J. (1976). Amino acid assignment to one of three blood–brain barrier amino acid carriers. *Am. J. Physiol.*, **230**, 94–98
18. Davson, H. and Pollay, M. (1963). The turnover of $^{24}$Na in the cerebrospinal fluid and its bearing on the blood–brain barrier. *J. Physiol.*, **167**, 247–255
19. Claude, P. and Goodenaugh, D.A. (1973). Fracture faces of zonula occludens from tight and leaky epithelia. *J. Cell Biol.*, **58**, 390–400
20. Connell, C.J. and Mercer, K.L. (1974). Freeze-fracture appearance of the capillary endothelium in the cerebral cortex of mouse brain. *Am. J. Anat.*, **140**, 595–598

21. Van Deurs, B. and Koehler, J.K. (1979). Tight junctions in the choroid plexus epithelium. *J. Cell Biol.*, **80**, 662–673
22. Bradbury, M.W.B. (1979). *The Concept of the Blood–Brain Barrier*. (Wiley, Chichester)
23. Michell, C.C. (1985). The Malpighi Lecture. Vascular permeability – the consequence of Malpighi's hypothesis. *Int. J. Microcirc. Clin. Exp.*, **4**, 265–284
24. Curry, F.E. and Michel, C.C. (1980). A fiber matrix model of capillary permeability. *Membrane Res.*, **20**, 96–99
25. Simionescu, N., Simionescu, M. and Palade, G.E. (1975). Permeability of muscle capillaries to small heme-peptides. Evidence for the existence of patent trans-endothelial channels. *J. Cell Biol.*, **64**, 586–607
26. Renkin, E.M. (1979). Multiple pathways of capillary permeability. *Circ. Res.*, **41**, 735–743
27. Bundgaard, M. (1980). Transport pathways in capillaries: in search of pores. *Ann. Rev. Physiol.*, **42**, 325–336
28. Frøkjaer-Jenson, J. (1983). The plasma luminal vesicular system in capillary endothelium conventional electron microscope (EM) thin sections compared with the picture arising from ultrathin (~140Å) serial sectionary. *Prog. Appl. Microcirc.*, **1**, 17–34
29. Müllgard, K., Balslev, Y. and Saunders, N.R. (1988). Structural aspects of the blood–brain and blood–CSF barriers with regard to endogenous proteins. In: *Peptide and Amino Acid Transport Mechanisms in the Central Nervous System*, eds., Lj. Rakic, D.J. Begley, H. Davson and B.V. Zlokovic, (Macmillan, London), pp. 93–102
30. Dewhurst, W.G. and Marley, E. (1965). Action of sympathomimetic and allied amines on the central nervous system of the chicken. *Br. J. Pharmacol.*, **25**, 671–727
31. Draskoci, M., Feldbury, W. and Haranath, P. (1968). Passage of circulating adrenalin into perfused cerebral ventricles and subarachnoid space. *J. Physiol.*, **150**, 24–49
32. Björklund, A., Falck, B., Hromek, F. and Owman, Ch. (1969). An enzyme barrier mechanism for monoamine precursors in newly-forming brain capillaries following electrolytic or mechanical lesions. *J. Neurochem.*, **16**, 1605–1608
33. Zlokovic, B.V., Begley, D.J., Segal, M.B., Davson, H., Rakic, Lj., Lipovac, M.N., Mitrovic, D.M. and Jankov, R.M. (1988). Neuropeptide transport mechanisms in the central nervous system. In: *Peptides and Amino Acid Mechanisms in the Central Nervous System*, eds., Lj. Rakic, D.J. Begley, H. Davson and B.V. Zlokovic, (Macmillan, London), pp. 3–20
34. Nabeshima, S., Rose, T.S., Landis, D.M.O. and Brightman, M.W. (1975). Junctions in the meninges and marginal glia. *J. Comp. Neurol.*, **164**, 127–170
35. Lorenzo, A.V. and Snodgrass, S.R. (1972). Leucine transport from ventricles and the cranial subarachnoid space in the cat. *J. Neurochem.*, **19**, 1287–1298
36. Wright, E.M. (1974). Active transport of glycine across the frog arachnoid membrane. *Brain Res.*, **76**, 354–358
37. Levin, E., Sepulveda, F.V. and Yudilevich, D.L. (1974). Pial vessel transport of substances from cerebrospinal fluid to blood. *Nature (London)*, **249**, 266–267
38. McComb, J.G., Davson, H. and Hollingsworth, J.R. (1975). Further studies on the difference between ventricular and subarachnoid perfusion. *Brain Res.*, **89**, 81–91
39. Kido, D.K., Gomez, D.G., Pavese, A.M. and Potts, D.G. (1976). Human spinal arachnoid granulations. *Neuroradiology*, **11**, 221–228
40. Jackson, R.T., Tigges, J. and Arnold, U. (1979). Subarachnoid space of the CNS and nasal mucosa and lymphatic system. *Arch. Otolaryngol.*, **150**, 180–184
41. Bradbury, M.W.B., Cserr, H.F. and Westrop, R.J. (1981). Drainage of cerebral interstitial fluid into deep cervical lymph of the rabbit. *Am. J. Physiol.*, **240**, F329–336
42. Brightman, M.W.B. and Reese, T.S. (1969). Junctions between intimately apposed cell membranes in the vertebrate brain. *J. Cell Biol.*, **40**, 648–677
43. Segal, M.B. and Burgess, A.M.C. (1974). A combined morphological study of the secretory processes in the rabbit choroid plexus. *J. Cell Sci.*, **14**, 339–350
44. Van Deurs, B.G. and Koehler, J.K. (1979). Tight junctions of the choroid plexus epithelia. *J. Cell Biol.*, **80**, 662–673

45. Brightman, M.W., Prescott, L. and Reese, T.S. (1975). Intercellular junctions of special ependyma. In: *Brain Endocrine Interaction*. II. *The Ventricular System*. (Karger, Basel), pp. 146–165
46. Welch, K. and Sadler, K. (1966). Permeability of the choroid plexus of the rabbit to several solutes. *Am. J. Physiol.*, **210**, 652–660
47. Deane, R. and Segal, M.B. (1985). The transport of sugars across the choroid plexus of the sheep. *J. Physiol.*, **362**, 245–260
48. Preston, J.E., Segal, M.B., Walley, G. and Zlokovic, B.V. (1989). The uptake of amino acids by the choroid plexus of the sheep. *J. Physiol.*, **408**, 31–43
49. Martinez-Palomo, A. and Erlig, D. (1975). Structure of tight junctions in epithelia with different permeability. *Proc. Natl. Acad. Sci. USA*, **72**, 4487–4491
50. Brightman, M.W. (1965). The distribution within the brain of ferritin injected into the cerebrospinal fluid compartments. *J. Cell Biol.*, **26**, 99–123
51. Baulden, T.W. and Krigman, H.R. (1975). Differential permeability of cerebral capillary and choroid plexus to lanthanum ion. *Brain Res.*, **99**, 444–448
52. Van Deurs, B., Moller, M. and Amtorp, O. (1978). Uptake of horseradish peroxidase for CSF into choroid plexus of the rat with special reference to transepithelial transport. *Cell Tissue Res.*, **187**, 215–234
53. Davson, H., Welch, K. and Segal, M.B. (1987). *The Physiology and Pathology of the Cerebrospinal Fluid*. (Churchill Livingstone, Edinburgh)
54. Reese, T.S. and Brightman, M.W. (1968). Similarity of structure of epithelia overlying fenestrated cerebral capillaries. *Anat. Res.*, **160**, 414
55. Weindl, A. and Joynt, R.J. (1972). Ultrastructure of the ventricular walls. *Arch Neurol.*, **26**, 420–427

# 2 Methods for the study of the functions of the blood–brain barrier

The blood–brain barrier is a complex multiple site structure with both active and passive elements. Any methods used to study such a system will always encounter a number of limitations and inbuilt errors. The recent developments in pharmacology, immunology and molecular biology have provided a variety of new techniques, which, in conjunction with the need for a greater understanding of brain pathology, have given a fresh impetus to the development of new methods to study the blood–brain barrier. However, it is essential that in the interpretation of data obtained from these new methods both the underlying limitations of the method and the complexity of the blood–brain barrier be always kept in mind. This is especially critical when such results are used to extrapolate from the *in vitro* experiment to the whole animal or the extrapolation of animal data to man. In addition, kinetic constants and parameters, which can be used to quantify barrier function and which have been derived by different experimental methods, should always be subject to the appropriate mathematical transformation so that the data can be compared between various methods and species.

The methods used to study the functions of the blood–brain barrier can be divided broadly into two main groups depending on whether the method uses the whole animal, i.e. *in vivo*, or *in vitro* where the barrier is simplified in some way. In this latter group, the most widely used *in vitro* preparation has been that of isolated cerebral capillaries harvested from homogenized brain. An alternative approach has been to grow these endothelial cells in tissue culture on a supporting mesh to form a pure 'barrier' *in vitro*. Even greater simplification has been achieved by fragmentating the endothelial cells of these capillaries and studying the characteristics of potential transport proteins in these cell wall fragments.

The *in vivo* studies may be further classified. Firstly, the main subdivision depends on the particular interface of the blood–brain barrier under study and on the direction of movement, e.g. blood to brain, blood to CSF, CSF to blood, etc. The molecule under study can be placed in the compartment under consideration and the interaction of the molecule studied either at the cellular interface of the barrier or on the movement across the barrier between compartments.

## *IN VIVO* TECHNIQUES

### Blood to brain techniques

The methods used to study the movement of molecules from blood to brain can be divided on the basis of the length of time during which the interaction has been studied, i.e. rapid, intermediate or long-term. The choice of method for a given molecule is highly dependent on its stability in blood and on the rate of its interaction with barrier, which emphasizes the need for a complementary approach, rather than the limitations of the different methods. Among the most frequently used rapid kinetic studies are the intracarotid injection technique of Oldendorf[1] and the paired-tracer indicator dilution method developed originally by Chinard for the lung[2] and applied to brain by Crone[3] and by Yudilevich[4]. For molecules which have a slow uptake and cannot be detected in the brain during single capillary passage through the cerebral circulation, different approaches have been developed. Most of these are based on the Davson's[5] classical steady-state kinetic analysis of solute entry into the brain, but deviate from it in certain ways, particularly in the calculation of permeability surface area product (PS) and the constants of saturable processes, $K_m$ and $V_{max}$, which can be derived from the PS-value. From the technical view point these long-term kinetic studies can be further classified on the basis of i.v. injection with single-brain sampling or multiple-brain sampling analysis, such as, for example, the method of Ohno *et al.*[6] and the experimental approaches of Gjedde[7], Patlak *et al.*[8], Blasberg *et al.*[9] and Baños *et al.*[10]. In contrast to these methods is the vascular brain perfusion technique of Zlokovic *et al.*[11], which can be used to keep the concentration constant in the perfusate of a slowly-penetrating compound for a sufficient length of time (up to 20 min) to characterize and quantitate the interaction of these molecules at the blood–brain barrier. The vascular brain perfusion method has also been applied to characterize the transfer of rapidly penetrating molecules (up to 1 min) using the technique of Takasato *et al.*[12].

The final most exciting method for the study of the entry characteristics from blood to brain is Syrota's use of positron emission tomography in man[13]. This relatively non-invasive method will yield much data of interest both from the clinical and experimental points of view.

### Blood/CSF methods

The most frequently used 'CSF-to-blood' technique without doubt is that of ventriculo-cisternal perfusion; this has been used in animals as diverse as the dogfish, rat and even man. Other methods include intracerebroventricular injection techniques and perfusion from the ventricles to the aqueduct of Sylvius. These 'blood to CSF' methods mostly use the classical steady-state blood level approach, with sampling of CSF from the ventricles, cisterna magna or lumbar sac and the calculation of CSF/plasma ratios. A more specific approach has involved sampling the nascent CSF from the surface of the oil covered choroid plexus *in vivo*, which was developed by Ames *et al.*[14] and by Welch *et al.*[15]. Miner and Reed[16] used an alternative method by enclosing the plexus in a capsule *in vivo*. The choroid plexus may be isolated further by the use of the *in vitro* isolated, blood perfused, choroid plexus of the sheep, a technique developed by Pollay *et al.*[17] and used extensively by Segal[18].

In the present volume it is not possible, due to space limitations, to provide a detailed and comprehensive description of all the above mentioned methods, and we have decided to focus our interest primarily on techniques employed in peptide and amino acid transport studies at the blood–brain barrier.

## *IN VITRO* TECHNIQUES

### Isolated cerebral microvessels

For more than 15 years the blood–brain barrier has been studied *in vitro* by means of isolated brain microvessels. Joó and Karnushina[19] developed the first micromethod for isolating cerebral microvessels, and since then a number of similar techniques have been used to separate the vessels from the rest of the brain tissue. Briefly, the isolation procedure can be summarized in two steps: (1) homogenization of the brain and/or gray matter with a loosely fitting rotating pestle (collagenase digestion may also be applied) and (2) separation of the vessels by either sieving through nylon nets and/or density gradient centrifugation, or passage through a column of glass beads. Alternatively during the separation phase these procedures may be combined.

The need to study the blood–brain barrier physiology and pathology *in vitro* arose from the fact that the barrier is difficult to study in isolation *in vivo* since it is located at the continuous endothelium of cerebral blood vessels. This special endothelium functions with respect to its permeability and electrophysiological characteristics, like a transporting epithelium[20]. However, experimental models used to study the physiology of epithelial tissues *in vitro*, such as the Ussing chamber for the frog skin, or isolated renal microperfused tubules, offer a much better possibility for transport research compared to isolated cerebral microvessels. In the above epithelia, continuity of the cellular layer remains preserved during the isolation procedure which allows the cellular barrier to be studied as a membrane separating two compartments, so that the net flux of ions and the electrical parameters of the epithelium may be investigated.

In spite of its limitations, the isolated microvessel preparation has proved particularly useful in the study of the biochemical properties of the cerebral vasculature, such as the lipid composition of the endothelial cells, the amino acid and carbohydrate composition of the basement membrane and the determination of major protein components[21]. In addition, a number of different enzymatic activities[22] and the presence of various classical neurotransmitters and enzymes involved in their metabolism[23] have been shown. Other enzymes involved in the metabolism of cyclic nucleotides, including adenylate cyclase, guanylate cyclase and cyclic phosphodiesterase[24], have also been demonstrated. The preparation has been of particular interest in monitoring receptor-mediated transcytotic exchanges of larger molecules[25] as well as being successfully applied in various receptor-binding studies[26]. This technique has proved to be a powerful tool for studying the molecular pathology of brain microvessels, including responses to cholesterol feeding and hypertension, effects of ischaemia and brain oedema, effects of lead intoxication[21,27] and, more recently, characterization of amyloid angiopathy in Alzheimer's disease[28].

Although this preparation is most useful for above-mentioned purposes, it has been suggested that it can be inappropriate or even misleading for others[29]. For example, most studies of transport with the preparation have examined the uptake of radiolabelled solutes into the suspended tissue fragments. These disrupted sections of microvessels are usually about 100 $\mu$m in length and may be open at their broken ends so that both luminal and abluminal sides can contribute to the measured cellular uptake or binding[29]. However, in some cases, as for example for Na/K-ATP-ase, the exact localization to the abluminal side of the endothelium has been demonstrated[30]. It has been computed on the basis of surface area–volume relationships that rapidly penetrating molecules, such as glucose and some amino acids, may reach a half-steady-state equilibrium very rapidly, within 5–10 s[31]. The estimated kinetic constants may be very different from those obtained *in vivo*, as well as varying between values for isolated microvessels prepared by different groups. Some technical difficulties reside in the difficulty of obtaining a relatively pure preparation of the capillaries. The isolated cerebral capillaries may be contaminated with small arterioles and venules, with their smooth muscle, as well as substantial astrocytic contamination, and this may be responsible for a relatively large variation in computed kinetic constants between different groups.

## Endothelial cells monolayer culture

Preparations of endothelial cells in culture seem to offer a better possibility for the study of transport features of the blood–brain barrier *in vitro*. It has been shown that these cells in tissue culture may form a cellular sheet, providing a preparation permitting the study of transcellular transport similar to those obtained by employing Ussing chambers for epithelial tissues. As has been demonstrated[22,32], the primary culture of endothelial cells displays some of the essential characteristics for the blood–brain barrier, such as the appearance of tight junctions between the cells and the absence of pinocytic vesicles. In the first phase, the procedure of growing endothelial cell monolayers involves the separation of cerebral capillaries from the brain, which is identical to the procedure for isolating microvessels. This is followed by an enzymatic treatment, with collagenase/dispase to achieve removal of the basement membrane, as well as to facilitate the release of contaminating pericytes and neuronal attachments. The microvessel endothelial cells are then suspended in culture medium and seeded onto the growth surfaces, where they attach and grow out from the original microvessel. It has been shown that endothelial cells retain the apparent ultrastructure of tight intercellular junctions despite enzymatic treatment, which has been confirmed by scanning electron microscopy and transmission electron microscopy. The cells exhibit alkaline phosphatase activity, and the presence of Factor VIII antigen and $\gamma$-glutamyl-transpeptidase, which are typical markers for the endothelial cells. To determine transendothelial transport the cells may be grown on collagen-coated, fibronectin-treated nylon mesh squares[33]. The nylon mesh squares can then be placed across an Ussing style chamber and when the endothelial cell monolayer has grown across the mesh it is possible to study bidirectional transfer. The usefulness of this preparation for transport studies is, however, very dependent on the integrity of the sheet and the absence of any small holes where the nylon mesh is uncovered by cells. To date no preparation has as yet been demonstrated which has true continuity with a high transendothelial electrical resistance equivalent to that found *in vivo*.

## Isolation of specific blood–brain barrier proteins

One of the most useful aspects of the technique for isolating cerebral microvessels is at the first stage in isolation, chemical characterization and identification of blood–brain barrier specific proteins. For example, $\gamma$-glutamyl-transpeptidase is specific for the blood–brain barrier and is not found in brain cells[34]. It has been suggested that other candidates for blood–brain barrier specific proteins include 50–55 kDa triplet of phosphoproteins and a series of low-molecular weight proteins in the range of 14–18 kDa, which are demonstrated in the plasma membrane fractions of endothelial cells, but not in synaptosomal plasma membrane fractions[34]. An important specific protein constituent is the 46 kDa protein asymmetrically localized on the lateral surface membrane of brain capillary endothelia, which has been demonstrated by the use of rabbit antiserum against bovine blood–brain barrier plasma membranes[35]. This antiserum selectively binds to the 46 kDa protein of the endothelial cell plasma membrane fraction, and does not bind to endothelia in other organs such as liver, heart or kidney. It has been suggested that it may represent a tight junctional protein[36].

Recently, the glucose transporter from microvessels has been separated from sheep brain[37], as well as from cerebral microvessels of the rat and pig[38] by means of a cytochalasin B binding, immunological reactivity and photoaffinity labelling technique. By using the same method, the distribution of glucose transporter in the mammalian brain was described[39], and the characterization of a nucleoside transporter in cerebral microvessels and the choroid plexus has been demonstrated[40].

## MATHEMATICAL ANALYSIS OF *IN VIVO* TECHNIQUES

### Davson's kinetic analysis of solute entry into the brain, cerebrospinal fluid and aqueous humour

Davson's classical kinetic analysis of solute penetration into the brain, cerebrospinal fluid and aqueous humour is based on a two-compartment 'permeability theory', one of the compartments being the blood plasma, and the other either in the brain and/or CSF or aqueous humour[41,42]. If the second compartment, let us say for example, the brain, is considered in relative isolation in terms that any significant net flow of test-solute through this compartment does not take place, then the concentration of test-solute in the brain, $C_{Br}$, may be related to the concentration in plasma, $C_{pl}$, and time of penetration, $t$, by two rate or fractional transfer constants, $k_{in}$ and $k_{out}$ as follows:

$$\frac{dC_{Br}}{dt} = k_{in}C_{pl} - k_{out}C_{Br} \qquad (1)$$

A similar equation may be applied in the case when the second compartment is CSF or aqueous humour. This approach is suitable for both the brain and the eye since the rate of entry is limited by the permeability of the barriers, which are represented by the continuous capillary endothelium for the brain and retina, or the continuous epithelium of the choroid plexus and ciliary processes when penetration into CSF or aqueous humour is considered. The units of the transfer constants are given in time, since the constants represent the fraction of the total volume of the compartment exchanged in unit time.

With substances which penetrate these barriers relatively slowly the back-flux may be ignored during the early stages of equilibration, in which case equation (1) simplifies to:

$$k_{in} = \frac{1}{C_{pl}} \cdot \frac{dC_{Br}}{dt} \qquad (2)$$

When equation (1) is integrated from zero time ($C_{Br} = 0$) to time $t$, it yields:

$$C_{Br} = R(1 - e^{-k_{out}t}) \qquad (3)$$

where $R$ is the ratio $k_{in}/k_{out}$ and/or the ratio $C_{Br}/C_{pl}$ at the time when equilibrium is reached. A numerical value for $k_{out}$ may be obtained from equation (3) as:

$$k_{out} = \frac{-\ln(R - C_{Br}/C_{pl})}{t} \qquad (4)$$

For solutes which penetrate rapidly from plasma into the brain compartment the two transfer constants are usually equal, $R = 1$, and the combined value can be computed from equation (4). The concentration of test-solute in plasma, $C_{pl}$, can be maintained at a constant level by continuous infusion at a diminishing rate, as recommended originally by Davson, or single i.v. injection when an integrated plasma concentration at time $t$ has to be computed. This technique is most suitable for the study of blood-to-brain or blood to CSF penetration of fast-penetrating non-electrolytes, inert polar molecules, slowly-penetrating or fast-penetrating non-electrolytes, inert polar molecules, slowly-penetrating ions as well as some drugs. However, these steady state i.v. methods are not suitable for the study of solutes which are subject to rapid metabolism by the liver, kidney, lung or in the blood compartment itself. To some extent the problem of metabolism may be diminished by surgical procedures such as portocaval anastomosis or nephrectomy, and use of more stable synthetic analogues of non-metabolizable compounds.

## Oldendorf's intracarotid injection technique

By means of Oldendorf's intracarotid injection technique it is possible to measure clearance of radiolabelled substances by brain during a single capillary passage, after an arterial injection using a diffusible internal standard[1,43,44]. Briefly, a 27-gauge needle or fine polythene tubing may be inserted into the right common carotid artery usually in the rat, and $^{14}C$-labelled test-substance with [$^{3}H$]water, as an internal standard, or $^{3}H$-labelled test-substance with [$^{14}C$]butanol or [$^{14}C$]antipyrine as an internal reference, is rapidly injected, and 5–15 s after the injection the animal is decapitated and brain taken for radioactivity analysis. The short time interval is chosen to prevent rapid back-flux of the reference tracer from brain tissue into the cerebral circulation. The percentage clearance by brain as obtained by this method is called a 'brain uptake index' (BUI) and it is calculated by employing the following equation:

$$\mathrm{BUI} = \frac{\text{brain}\,^{14}\text{C} - \text{test substance/brain}\,^{3}\text{HOH}}{\text{injected}\,^{14}\text{C} - \text{test substance/injected}\,^{3}\text{HOH}} \times 100 \qquad (5)$$

$$\mathrm{BUI} = \frac{\text{brain}\,^{3}\text{H} - \text{test substance/brain}\,^{14}\text{C reference}}{\text{injected}\,^{3}\text{H} - \text{test substance/injected}\,^{14}\text{C reference}} \times 100 \qquad (6)$$

(All the values are expressed as dpm/g of tissue or dpm/ml of injectate)

Subsequently, the method has been further developed to calculate an actual extraction fraction ($E$) by correcting for the incomplete extraction of the reference substance ($E_{HOH}$, for example), as well as by applying an $^{113}$Indium correction for the residual unwashed radioactivity in the blood by means of following equation when $^{3}$HOH is used as a reference:

$$E = E_{HOH}\left[\frac{\text{brain}\,^{14}\text{C/brain}\,^{3}\text{H}}{\text{injected}\,^{14}\text{C/injected}\,^{3}\text{H}} - \frac{\text{brain}\,^{113}\text{In/brain}\,^{3}\text{H}}{\text{injected}\,^{113}\text{In/injected}\,^{3}\text{H}}\right] \times 100 \qquad (7)$$

In a similar manner, the BUI value for the test molecule may be corrected for the residual vascular radioactivity for BUI of sucrose which remains in the cerebral vessels. Thus the corrected BUI value may be obtained by equation[45,46]:

$$\mathrm{BUI}_{corr.} = \mathrm{BUI}_{test} - \mathrm{BUI}_{sucrose} \qquad (8)$$

If it is assumed that injectate totally fills the cerebral capillaries at the time of decapitation, the concentration of test substance in the plasma, $C_{pl}$, is equal to the injectate concentration $C_i$, and, according to the carrier theory at the blood–brain barrier, the BUI value is related to Michaelis–Menten kinetic parameters, $K_m$, $V_{max}$ and $K_d$ by means of the following equations, as developed by Pardridge and Oldendorf[47] and Pardridge and Meitus[45], for BUI results:

$$\frac{J_{in}}{\overline{C}} = EF = \frac{V_{max}}{K_m + \overline{C}} + K_d \qquad (9)$$

where $J_{in}$ is the unidirectional influx, $E$ is the extraction derived from the BUI value either by using equation (7) or the $\mathrm{BUI}_{corr.}$ value, equation (8), corrected by the $E_{HOH}$ by the extraction of a diffusible internal standard; $F$ is cerebral blood flow. $\overline{C}$ is the mean capillary concentration and in general $C_i$ equals $\overline{C}$ when $E<35\%$. However, since E may exceed 35% in different transport studies it is better to calculate the mean capillary concentration from the $C_i$ value by using the equation[48]:

$$\overline{C} = \frac{E}{-\ln(1-E)} \cdot C_i \qquad (10)$$

Extraction is related to the permeability surface area product, *PS*, by Crone's equation:

$$PS = -F\ln(1-E) \qquad (11)$$

where *P* is the cerebrovascular permeability coefficient and *S* is the surface area of capillaries in the brain, assumed in mammals, to be about 100 $cm^2g^{-1}$ brain tissue[49].

The Oldendorf method is most useful in characterizing rapidly penetrating solutes from blood-to-brain, and a wide range of substrates has been studied including amino acids, amines, hexoses, inert polar substances, organic acids, nucleic acid precursors, peptides, different drugs and, more recently, plasma protein mediated transport. Also full kinetic characterization of different blood–brain barrier transport systems *in vivo* and the effects of anaesthesia have been performed by means of this method. However, for slowly-penetrating compounds the technique reaches its limit of sensitivity and the type of interaction of such compounds with a blood–brain barrier is difficult to characterize. For example, although this method reveals that ions, peptides, proteins and slowly-penetrating inert polar molecules do not penetrate rapidly from blood into the brain, a further characterization of their interactions with the blood–brain barrier is not possible since all these molecules exhibit a negligible or zero extraction (*E*) during single passage through the cerebral circulation. Recently, the Oldendorf method has been modified and applied to the dogfish. This technique offers a unique opportunity for studying glial physiology *in vivo* and the permeability of the glial blood–brain barrier in this species to amino acids, sugars and other rapidly penetrating compounds[50].

## Indicator diffusion technique

Brain uptake of rapidly penetrating molecules can be studied also by means of the 'indicator diffusion' method, originally used for measurements of capillary permeability by Chinard *et al.*[2]. The method has been subsequently applied to study the transport of a number of non-electrolytes including glucose across the blood–brain barrier by Crone[51,52] and later to the uptake of glucose and amino acids by Yudilevich *et al.*[4,53]. When the indicator diffusion method is applied to the brain, a single arterial injection of radiolabelled test molecule together with a non-diffusible reference tracer is made,

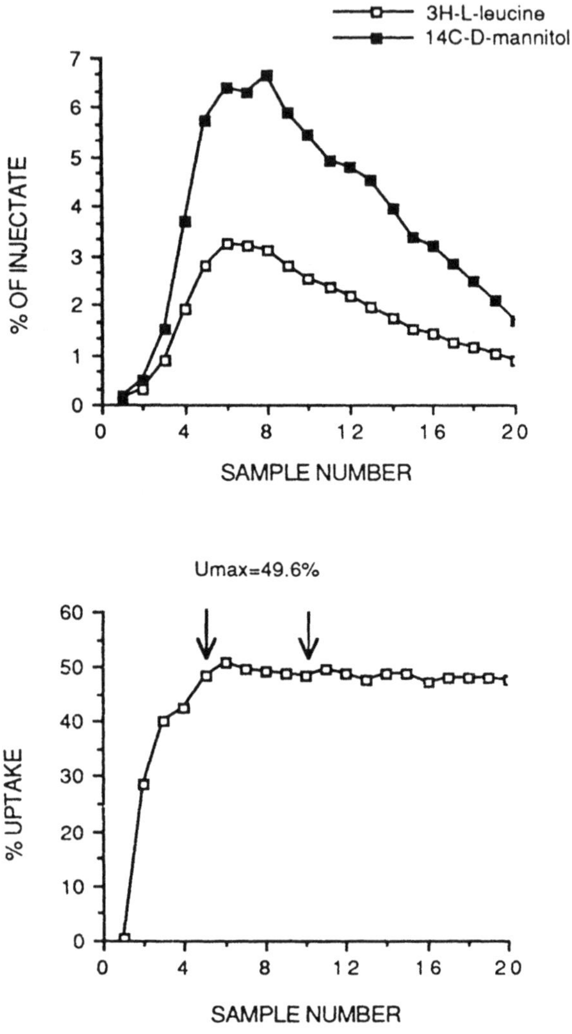


**Figure 2.1 Upper:** typical dilution curves obtained after simultaneous arterial injection of [$^{3}$H]L-leucine and [$^{14}$C]D-mannitol in the isolated perfused sheep choroid plexus. Twenty sequential venous effluent samples are collected after injection and the tracer concentrations in each sample expressed as a percentage of their respective injected doses; **Lower:** uptake of [$^{3}$H]L-leucine relative to [$^{14}$C]D-mannitol (uptake % = {1–(% injectate $^{3}$H/% injectate $^{14}$C)} x 100) in successive venous samples is plotted. Initial maximal uptake ($U_{max}$) is calculated as the mean of those uptake values between the arrows and corresponds to peak recovery of $^{3}$H and $^{14}$C. Taken from Preston *et al.* (1989)[66]

and samples of cerebral venous drainage are collected at short time intervals, usually at the confluence of the sinuses, i.e. from the torcular blood. In order to estimate the fractional extraction of the test molecule into the brain, the radioactivities of both tracers in the venous samples are compared using the following equation:

$$E = 1 - \frac{\text{radioactivity of test molecule}}{\text{radioactivity of reference molecule}} \qquad (12)$$

From the fractional extraction vs. time plot later called the uptake curve (Figure 2.1) (see for review, Yudilevich *et al.*[54]), the plateau area can be defined as the maximal cellular uptake ($U_{max}$) which is related to the permeability surface area, the product of $P$ and $S$ by Crone's equation:

$$PS = -F \ln (1 - U_{max}) \qquad (13)$$

where $F$ is the cerebral blood flow, and $U_{max}$ the unidirectional uptake across the blood–brain barrier.

Under appropriate experimental conditions, by varying the test substance concentration, Michaelis–Menten kinetic parameters can be obtained by using the equation:

$$PS = \frac{U_{max}}{C_{pl} + K_m} + K_d \qquad (14)$$

Kinetics of unidirectional influx based on $U_{max}$ measurements were made first for glucose[55] and later for amino acids [56]. Recently, the indication diffusion method, now called the 'paired-tracer indicator dilution technique', has been applied by Zlokovic *et al.*[57] to the the perfused choroid plexus *in situ* to characterize cellular uptake of peptides and amino acids at the blood–tissue interface of the choroidal epithelium[58] (Figure 2.1). However, this method cannot be used for studying the blood–brain barrier transport of compounds which are subject to metabolic transformation while passing through the cerebral circulation, or which permeate slowly through the blood–brain barrier. With respect to this point, the paired-tracer indicator diffusion technique is more similar to the intracarotid injection method of Oldendorf, than to the brain sampling techniques after intravenous injection of test solute.

## Brain sampling techniques after intravenous injection of test solute

A number of techniques of brain sampling followed by direct measurement of radioactivity in the brain after i.v. injection or infusion of test solute with measured plasma concentration have been developed. In some of them the brain is sampled within 3 min or less after a constant arterial blood concentration of test solute has been established by means of a rapid intravenous injection at an exponentially decaying rate[10,59]. A mean capillary concentration of test solute should be used and calculated in these experiments rather than the arterial concentration, particularly for fast-exchanging solutes. Also, when this method is used for the characterization of absolute kinetic constant of a carrier system, it is important to bear in mind that the cerebral capillaries contain the animal's own blood during the period of exchange, and these endogenous substrates will compete with those under study and so alter the derived kinetic constants. It seems most likely that this method will not be applicable to the study of fast-exchanging substances, since a rapid backflux from the brain tissue into the blood can occur.

A method which is suitable for the measurement of the lower limits of cerebrovascular permeability to non-electrolytes has been developed by Ohno *et al.*[6]. Both the artery and vein in the anaesthetized animal are cannulated, and following a rapid i.v. injection of test solute, arterial blood is collected at regular time intervals so that an integrated arterial concentration at the time, $T$, of brain sampling can be calculated. The permeability surface area product by this technique may be computed from the following equation:

$$PS = \frac{C_{Br}}{\int_0^T C_{pl}\, dt} \tag{15}$$

The method has proved to be useful for slowly-penetrating non-metabolizable compounds. However for molecules that are metabolized in the liver, kidney, lung or blood itself, there is uncertainty as to whether the label is still representative of the intact molecule.

The advantage of the intravenous bolus administration of labelled test solute to determine the unidirectional transfer constant, $K_{in}$, for blood to brain, from multiple-time brain uptake data has been recognized by Gjedde[60,61] and by Patlak *et al.*[62]. This type of analysis includes the possibility of calculating the initial distribution of the test solute in rapidly reversible compartments (vascular compartment and/or compartments in parallel with the blood–brain barrier) by means of the equation:

$$\frac{C_{Br}(T)}{C_{pl}(T)} = \frac{K_{in} \int_0^T C_{pl}(t)\,dt}{C_{pl}(T)} + V_i \tag{16}$$

where $C_{Br}(T)$ is the total amount of test solute measured per unit mass of brain tissue at time $T$, $C_{pl}(T)$ is the concentration of test solute in the plasma at time $T$, and $V_i$ is the initial volume of distribution of the test-solute. This equation defines a straight line with slope $K_{in}$ (the unidirectional transfer constant) and ordinate intercept, $V_i$. The initial rate constant for unidirectional transfer ($K_{in}$) can also be determined using 'simple two-compartment analysis' and a special case when the backflux (brain to blood) is much smaller than the influx (blood to brain) during the experimental period[8], namely:

$$K_{in} = \frac{C_{Br}}{\int_0^T C_{pl}\,dt} \tag{17}$$

that is similar to equation (15). The $K_{in}$ value is given in units ml min$^{-1}$ g$^{-1}$ and represents in fact the experimentally determined *PS* product. Limitations of multiple brain uptake analysis for studying highly metabolizable molecules after i.v. injection are the same as those enumerated above for other i.v. methods.

## Vascular brain perfusion methods

Perfusion of the head with an artificial medium is by no means a new technique and has been employed by a number of authors, in the rat[12,63,64] and in the mouse[65], but in each case it was not certain whether the technique provided an adequate oxygenation of the brain, so permitting the maintenance of an intact blood–brain barrier and normal water and electrolyte balance. Recently, Zlokovic *et al.*[11] have developed the method of vascular perfusion of the ipsilateral forebrain of the guinea pig, with a perfusate containing washed sheep red cells. The physiological parameters such as EEG, ECG and respiratory movements, and biochemical parameters such as level of ATP, ECP and lactate, water and electrolyte content, cerebral blood flow and the barrier to mannitol, all remained apparently normal for at least 30 min of perfusion (Figure 2.2). Graphical multiple-time brain uptake analysis and equation (16) have been most frequently employed for characterizing the interaction of test solute with the barrier. Since under perfusion conditions the concentration of test solute in the arterial inflow remains constant, i.e.

$$\int_0^T C_{pl}(t)\,dt = C_{pl}(T)$$

the equation becomes:

$$\frac{C_{Br}(T)}{C_{pl}(T)} = K_{in}T + V_i \quad (18)$$

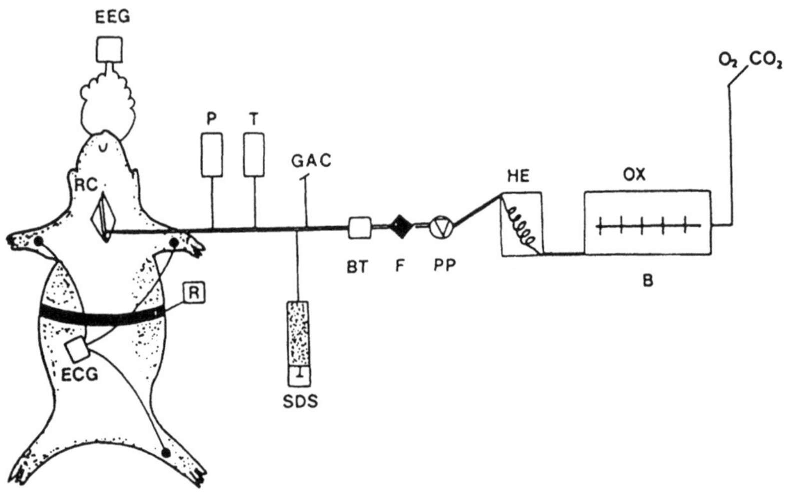


**Figure 2.2** A line diagram of the perfusion circuit. OX, oxygenator; B, blood reservoir; HE, heat exchanger; PP, peristaltic pump; F, filter; BT, bubble trap; GAC, gas analysis control; SDS, slow drive syringe; T, temperature control; P, pressure transducer; EEG, platinum electrodes for encephalogram; ECG, siver disc electrodes for electrocardiogram; R, respiratory belt; RC, position of cannula into right common carotid artery in the guinea pig. Zlokovic *et al.* (1986)[11]

If the single sampling method is applied, then a correction for the initial volume of distribution should be made to give a more accurate value for $K_{in}$ and the equation becomes:

$$K_{in} = \frac{[C_{Br}(T)/C_{pl}(T)] - V_i}{T} \qquad (19)$$

As $K_{in}$ represents actually the permeability surface area products, according to carrier theory:

$$K_{in} = \frac{V_{max}}{K_m + C_{pl}} + K_d \qquad (20)$$

or:

$$J_{in} = \frac{V_{max} \cdot C_{pl}}{K_m + C_{pl}} + K_d \cdot C_{pl} \qquad (21)$$

where $J_{in}$ is the unidirectional influx, and $V_{max}$, $K_m$ and $K_d$ are Michaelis–Menten constants.

The vascular brain perfusion method of Zlokovic has been most useful in characterizing the uptake of molecules which interact slowly with the luminal side of the blood–brain barrier, and so represents an approach which complements the method of Oldendorf. The method has proved to be useful in particular for studying the interactions of blood-borne molecules such as peptides and proteins with the blood–brain barrier since these molecules are often subject to metabolic transformation, by the blood or by the liver and lung. In addition the true physiological concentration of such compounds can be used and the method combined with immunocytochemical and autoradiographic techniques. The vascular perfusion method is not suitable for studying the transport of rapidly-penetrating solutes owing to their high flow-dependency and their rapid backflux from the brain tissue into the blood.

## REFERENCES


1. Oldendorf, W.M. (1970). Measurement of brain uptake of radiolabelled substances using a tritiated water internal standard. *Brain Res.*, **24**, 372–376
2. Chinard, F.P., Vosburgh, G.J. and Enns, T. (1955). Transcapillary exchange of water and of other substances in certain organs of the dog. *Am. J. Physiol.*, **183**, 221–234
3. Crone, C. (1961). Gu diffusionen af nogle organisk non-elektrolyter fra clod til hjeznevaeu. (Munksgaard, Copenhagen)
4. Yudelivich, D.L., De Rose, N. and Sepulveda, F.V. (1972). Facilitated transport of amino acids through the blood–brain barrier of the dog studied in a single capillary circulation. *Brain Res.*, **44**, 569–578
5. Davson, H. (1955). A comparative study of the aqueous humour and cerebrospinal fluid in the rabbit. *J. Physiol.*, **129**, 111–133
6. Ohno, K., Pettigrew, K.D. and Rapoport, S.I. (1978). Lower limits of cerebrovascular permeability to nonelectrolytes in the concious rat. *Am. J. Physiol.*, **235**, H299–H307
7. Gjedde, A. (1981). High and low-affinity transport of *D*-glucose from blood to brain. *J. Neurochem.*, **36**, 1463–1471
8. Patlak, C.S., Fenstermacher, J.D. and Blasberg, R.G. (1983). Graphical evaluation of blood-to brain transfer constants from multiple-time uptake data. *J. Cereb. Blood Flow Metab.*, **3**, 1–7
9. Blasberg, R.G., Fenstermacher, J.D. and Patlak, C.S. (1983). Transport of gamma-aminoisobutyric acid across brain capillary and cellular membranes. *J. Cereb. Blood Flow Metab.*, **3**, 8–32
10. Baños, G., Daniel, P.M. and Pratt, O.E. (1974). Saturation of shared mechanism which transports *L*-arginine and *L*-lysine into the brain of the living rat. *J. Physiol.*, **236**, 29–41
11. Zlokovic, B.V., Begley, D.J., Duricic, B.M. and Mitrovic, D.M. (1986). Measurement of solute transport across the blood–brain barrier in the perfused guinea-pig brain, method and application to N-methyl-aminoisobutyric acid. *J. Neurochem.*, **46**, 1444–1451
12. Takasato, Y., Rapoport, S.I. and Smith, Q.R. (1984). An *in situ* brain perfusion technique to study cerebrovascular transport in the rat. *Am. J. Physiol.*, **247**, H484–H493
13. Syrota, A. (1985). Solute transport and ligand–receptor interactions studied in man by positron-emission tomography. In: *Carrier-Mediated Transport of Solutes from Blood to Tissue*, eds., D.L. Yudilevich and G.E. Mann (Longman, New York), pp. 155–176
14. Ames, A., Higashi, U.K. and Nesbett, F.B. (1965). Effects of $PCO_2$ acetazolamide and ouabain on volume and composition of choroid plexus fluid. *J. Physiol.*, **181**, 516–524
15. Welch, K. and Sadler, K. (1966). Permeability of the choroid plexus of the rabbit to several solutes. *Am. J. Physiol.*, **210**, 652–660

16. Miner, L.C. and Reed, D.J. (1972). Composition of fluid obtained from choroid plexus tissue isolated in a chamber *in situ*. *J. Physiol.*, **227**, 127–139
17. Pollay, M., Stevens, A., Estrada, E. and Kaplan, R. (1972). Extracorporeal perfusion of choroid plexus. *J. Appl. Physiol.*, **32**, 612–617
18. Deane, R. and Segal, M.B. (1985). The transport of sugars across the perfused choroid plexus of the sheep. *J. Physiol.*, **362**, 245–260
19. Joo, F. and Karnushina, I. (1973). A procedure for the isolation of capillaries from rat brain. *Cytobios.*, **8**, 41–48
20. Davson, H. (1970). *A Text Book of General Physiology*. (Churchill Livingstone, London)
21. Joó, F. (1985). The blood–brain barrier *in vitro*: ten years of research on microvessels isolated from the brain. *Neurochem. Int.*, **7 (1)**, pp. 1–25
22. Spatz, M. and Mrsulja, B.B. (1982). Progress in cerebral microvascular studies related to the function of the blood–brain barrier. *Adv. Cell Neurobiol.*, **3**, 311–337
23. Duricic, B. and Mrsulja, B.B. (1988). Transport and barrier systems of the cerebral microvasculature: enzymatic aspects. In: *Peptide and Amino Acid Transport Mechanisms in the Central Nervous System*, eds., L.J. Rakic, Lj. Begley, H. Davson and B.V. Zlokovic (Macmillan, London) pp. 269–279
24. Joó, F. (1988). Cyclic nucleotide-mediated regulation of albumin transport in brain microvessels. In: *Peptide and Amino Acid Transport Mechanisms in the Central Nervous System*, eds., L.J. Rakic, D.J. Begley, H. Davson and B.V. Zlokovic (Macmillan, London) pp. 119–129
25. Pardridge, W.M. (1988). Recent advances in blood–brain barrier transport. *Anu. Rev. Pharmacol. Toxicol.*, **28**, 25–39
26. Betz, A.L. and Goldstein, G.W. (1986). Specialized properties and solute transport in brain capillaries. *Anu. Rev. Physiol.*, **48**, 241–250
27. Joó, F. (1986). New aspects to the function of cerebral endothelium. *Nature*, **321**, 197–198
28. Pardridge, W.M., Vinters, H.V., Yang, J., Eisenberg, J., Choi, T.B., Tourtelotte, W.W., Huebuer, V. and Shirely, J.E. (1987). Amyloid angiopathy of Alzheimer's disease: amino acid composition and partial sequence of a 4,200-dalton peptide isolated from cortical microvessels. *J. Neurochem.*, **49**, 1394–1401
29. Bradbury, M.W. (1985). Critique. In: The blood–brain barrier *in vitro*: ten years of research on microvessels isolated from the brain. *Neurochem. Int.*, **Vol. 7**, pp. 27–28
30. Goldstein, G.W. and Betz, A.L. (1986). The blood–brain barrier. It is not an impassable wall but a highly selective gate that bars some substances and admits others. *Sci. Am.*, **255**, 74–83
31. Bowman, P.D., Betz, A.L., Wolminsky, J.S., Penney, J.B., Shivers, R.R. and Goldstein, G.W. (1981). Primary culture of capillary endothelium from rat brain. *In Vitro*, **17**, 353–362
32. Bowman, P.D., Ennis, S.R., Razey, K.E., Betz, A.L. and Goldstein, G.W. (1983). Brain microvessels endothelial cells in tissue culture: a model for study of blood–brain barrier permeability. *Am. Neurol.*, **14**, 396–402
33. Audus, K.L. and Borchardt, R.T. (1986). Characterization of an *in vitro* blood–barrier model system for studying drug transport and metabolism. *Pharmaceut. Res.*, **3 (2)**, 31–87
34. Pardridge, W.M., Yang, J. and Eisenberg, J. (1985). Blood–brain barrier protein phosphorylation and dephosphorylation. *J. Neurochem.*, **45**, 1141–1147
35. Pardridge, W.M., Yang, J., Eisenberg, J. and Mietus, L.J. (1986). Antibodies to blood–brain barrier bind selectively to brain capillary endothelial lateral membranes and to a 46K protein. *J. Cereb. Blood Flow Metab.*, **6**, 203–211
36. Pardridge, W.M. (1986). Mechanisms of neuropeptide interaction with the blood–brain barrier. *Ann. N.Y. Acad. Sci.*, **481**, 231–249
37. Baldwin, S.A., Brewster, F., Cairns, M.T., Gardiner, R.M. and Ruggier, R. (1984). Identification of *D*-glucose sensitive cytochalasin B binding component in isolated ovine cerebral microvessels. *J. Physiol.*, 82(P)

38. Dick, A.P.K., Harik, S.I., Klip, A. and Wolmer, D.M. (1984). Identification and characterization of the glucose transporter of the blood–brain barrier by cytochalasin B binding and immunological reactivity. *Proc. Natl. Acad. Sci. USA*, **81**, 7233–7237
39. Dick, A.P.K. and Harik, S.I. (1986). Distribution of the glucose transporter in the mammalian brain. *J. Neurochem.*, **46**, 1406–1411
40. Kaloria, R.N. and Harik, S.I. (1986). Nucleoside transporter of cerebral microvessels and choroid plexus. *J. Neurochem.*, **47**, 1849–1856
41. Davson, H. (1955). A comparative study of the aqueous humour and cerebrospinal fluid in the rabbit. *J. Physiol.*, **129**, 111–133
42. Davson, H. and Matchett, P.A. (1953). The kinetics of penetration of the blood-aqueous barrier. *J. Physiol.*, **122**, 11–32
43. Oldendorf, W.H. and Szabo, J. (1976). Amino acid assignment to one of three blood–brain barrier amino acid carriers. *Am. J. Physiol.*, **230**, 94–98
44. Oldendorf, W.H. (1981). Clearance of radiolabelled substances by brain after arterial injection using a diffusible internal standard. In: *Research Methods in Neurobiochemistry*, Vol. 5, eds., N. Marlis and R. Rodright (Plenum Publishing Co., New York) pp. 91–112
45. Pardridge, W.M. and Mietus, L.J. (1982). Kinetics of neutral amino acid transport through the blood–brain barrier of the newborn rabbit. *J. Neurochem.*, **38**, 955–962
46. Zlokovic, B.V., Begley, D.J. and Chain, D.G. (1983). Blood–brain barrier permeability to dipeptides and their constituent amino acids. *Brain Res.*, **271**, 65–71
47. Pardridge, W.M. and Oldendorf, W.H. (1975). Kinetics of blood–brain barrier transport of hexoses. *Biochim. Biophys. Acta*, **382**, 377–392
48. Pappenheimer, J.R. and Setchell, B.P. (1973). Cerebral glucose transport and oxygen consumption in sheep and rabbits. *J. Physiol.*, **233**, 529–551
49. Bradbury, M.W. (1979). *The Concept of the Blood–Brain Barrier.* (Wiley, Chichester)
50. Abbott, N.J. and Zlokovic, B.V. (1987). A preparation for the study of glial transport: the glial blood–brain barrier of the dogfish *Scyliorhinus canicula*. *J. Physiol.*, **394**, 91
51. Crone, C. (1965). The permeability of brain capillaries to non-electrolytes. *Acta Physiol. Scand.*, **64**, 407–417
52. Crone, C. (1965). Facilitated transfer of glucose from blood into brain. *J. Physiol.*, **181**, 103–113
53. Yudilevich, D.L. and De Rose, N. (1971). Blood–brain transfer of glucose and other molecules measured by rapid indicator dilution. *Am. J. Physiol.*, **220**, 841–846
54. Yudilevich, D.L., Wheeler, C.P.D. and Bustamante, J.C. (1988). A comparative view on amino acid transport across the blood–brain barrier (endothelium) and the placenta (trophoblast). In: *Peptide and Amino Acid Transport Mechanisms in the Central Nervous System*, eds., L.J. Rakic, D.J. Begley, H. Davson and B.V. Zlokovic. (Macmillan, London) pp. 245–267
55. Betz, A.L., Gilboe, D.D., Yudilevich, D.L. and Drewes, L.R. (1973). Kinetics of unidirectional glucose transport into the isolated dog brain. *Am. J. Physiol.*, **225**, 586–592
56. Betz, A.L., Gilboe, D.D. and Drewes, L.R. (1975). Kinetics of unidirectional leucine transport into the brain: effects of isoleucine, valine and anoxia. *Am. J. Physiol.*, **228**, 895–900
57. Zlokovic, B.V., Segal, M.B., Begley, D.J., Davson, H. and Rakic, L.J. (1985). Permeability of the blood–cerebrospinal fluid and blood–brain barrier to thyrotropin-releasing hormone. *Brain Res.*, **358**, 191–199
58. Zlokovic, B.V., Segal, M.B., Davson, H. and Mitrovic, D.M. (1988). Unidirectional uptake of enkephalins at the blood–tissue interface of the blood–cerebrospinal fluid barrier: a saturable mechanism. *Regul. Peptides*, **20**, 33–44
59. Bachelard, H.S., Daniel, P.M., Love, E.R. and Pratt, O.E. (1973). The transport of glucose into the brain of the rat *in vivo*. *Proc. R. Soc. B.*, **183**, 71–82
60. Gjedde, A. (1982). Calculation of cerebral glucose phosphorylation from brain uptake of glucose analogues *in vivo*: a re-examination. *Brain Res. Rev.*, **4**, 237–274

61. Gjedde, A. (1988). Exchange diffusion of large neutral amino acids between blood and brain. In: *Peptide and Amino Acid Transport Mechanisms in the Central Nervous System*, eds., Lj. Rakic, D.J. Begley, H. Davson and B.V. Zlokovic (Macmillan, London) pp. 195–209
62. Patlak, C.S., Fenstermacher, J.D. and Blasberg, R.G. (1983). Graphical evaluation of blood to brain transfer constants from multiple time uptake data. *J. Cereb. Blood Flow Metab.*, **3**, 1–7
63. Andjus, R.K., Suhara, K. and Slosster, H.A. (1967). An isolated perfused rat brain preparation, its spontaneous and stimulated activity. *J. Appl. Physiol.*, **22**, 1033–1039
64. Thompson, A.M., Robertson, R.C. and Bauer, T.A. (1968). A rat head perfusion technique developed for the study of brain uptake of materials. *J. Appl. Physiol.*, **24**, 407–411
65. Bradbury, M.W., Deane, R. and Rosenberg, G. (1984). Regional blood flow, EEG and electrolytes in mouse brain perfused with perfluorochemical, FC43. *J. Physiol.*, **355**, 31
66. Preston, J.E., Segal, MB, Walley, G.T. and Zlokovic, B.V. (1989). Neutral amino acid uptake by the isolated perfused sheep choroid plexus. *J. Physiol.*, 408, 31–43

# 3 Amino acids and the blood–brain barrier

Amino acids are the structural element of all protein, so their entry from blood into brain is essential for brain growth and metabolism. However, these molecules can have an individual role as powerful neurotransmitters and their level in the brain ISF must be carefully regulated. This homeostasis has several possible sites for control, the entry step from the blood, the uptake and loss from the neurons and glia and the various drainage processes, either by diffusion into the CSF or back across the blood–brain barrier.

The exchange of amino acids between blood and brain can occur at three locations. Firstly, there is the large interface of the cerebral capillary endothelium, secondly the arachnoid membranes and thirdly the choroid plexuses and CSF. As stated by Bradbury in his book on the blood–brain barrier, 'the unravelling of the mechanisms for amino-acid transport at the blood–brain barrier is proving a complex, but engaging problem[1]. The reasons for these difficulties centre on the large number of amino acids, their similar physicochemical properties, which produce overlap in their affinities for the various classes of carrier molecule, species variation and the complexity of the blood–brain barrier.

As stated in Chapter 2, the methods available to study amino acid exchange between blood and brain are either *in vivo*, where it is difficult to limit the study to a single exchange interface, or *in vitro*, where the exchange interface is simplified, but the blood flow is lost and the validity of the model may be questioned. However, the *in vivo* steady-state methods using non-labelled amino acids have shown that the ability of the various classes of amino acids to enter the brain was different. Anionic amino acids (acidic) such as aspartate (ASP) and glutamate (GLU) seemed to be excluded from the brain whereas amino acids, such as arginine (ARG), histidine (HIS) and methionine (MET) could cross into the brain[2]. In fact tyrosine (TYR) entered brain as fast as it entered muscle, providing it was the L and not the D-isomer. The same was shown for tryptophan (TRY) and elevating the level of the D-isomer did not affect the entry of the L-form. Competition between amino acids for entry into the brain was shown between TYR and phenylalanine

(PHE), but not between alanine (ALA), arginine (ARG), GLU, glutamine (GLN), serine (SER) and threonine (THR)[3]. This was a key observation, since it could now be seen how the elevation of one amino acid in the blood could depress the entry into the brain of another amino acid, providing it was of a specific group and would thus lead to a specific brain deficiency of that amino acid[4]. Lajtha, Underfriend and others confirmed that the entry of amino acids into the brain was stereospecific, selective and saturable, so it must be a carrier-mediated process[5,6]. There is, however, a problem with steady-state type methods, when using labelled amino acids, since these molecules can be rapidly metabolized and the label detached. If the relationship between the label and the amino acid is not checked by chromatographic methods it is easy to be mislead. To avoid this problem, and to study the large number of amino acids present in the plasma, two rapid uptake methods were developed at about the same time. Yudilevich used Crone's single pass method which was originally developed by Chinard for the lung[7–9]. In this technique, a bolus is injected into the carotid artery and the venous effluent collected dropwise from the torcular blood (now called paired tracer indicator diffusion method). The uptake of the various amino acids are compared to that of a vascular marker which does not cross the BBB during a single transit of the brain. This method only determines uptake on the blood-side of the BBB, but the technique demonstrated that the uptake of some amino acids was carrier-mediated, and stereospecific, etc. Large neutral amino acids and HIS had a considerable uptake whereas small neutrals such as ALA and glycine (GLY) were not taken up, and nor were the cationic or anionic amino acids[9–11].

During the same period Oldendorf had developed his brain uptake index (BUI) technique in the rat. This method uses single observations in many animals and measures the uptake of amino acid relevant to that of $^{3}$HOH or a lipid soluble diffusible marker such as [$^{14}$C]butanol or [$^{14}$C]antipyrine. These initial studies with the BUI method showed some differences to the paired tracer method which may be related to a species variation or difference in methodology. As before, there was a high BUI for large neutral amino acids and HIS and, apart from TYR, the amino acids with the highest BUI were essential to the rat brain. However in addition, the BUI method revealed an uptake for the smaller neutrals and the basic amino acids. In later studies, where the precision of the method was improved, Oldendorf and his colleagues were able to demonstrate a low BUI for acidic amino acids and were able to assign most amino acids to one of three classes, each carried by its own carrier type, i.e. neutral, basic (cationic) and acidic (anionic). There was, however, little uptake for GLY, ALA or proline (PRO)[12,13].

The basic classification of amino acid carrier types in cell membranes has been due to the pioneering work of Christensen, who was also one of the first to study the entry of amino acids from blood into the CSF[14,15]. His work has laid the foundation for a system of classification, which is slowly evolving to include data from a variety of cell types and tissues. Table 3.1 shows a

**Table 3.1** Abbreviated summary of some known amino acid transport systms in tissues and cells of higher animal. According to Christensen and Kilberg (1987)[15]

| | | |
|---|---|---|
| **Systems for dipolar amino acids** | | |
| **$Na^+$-dependent systems** | | |
| System Gly | Glycine and sarcosine<br>Widespread | Characteristic tolerance of an N-methyl group |
| System A | Most dipolar amino acids<br>Ubiquitous | |
| System ASC | Numerous dipolar amino acids but not their N-methyl derivatives. At lowered pH, certain anionic amino acids | |
| System $B^{o,+}$ | Especially wide in scope | |
| System N | Histidine, glutamine, asparagine<br>Hepatocytes, distribution otherwise not determined | |
| $\beta$-system | $\beta$-alanine, taurine, $\beta$-aminobutyric acid | |
| **$Na^+$-independent systems** | | |
| System L* | Prefers dipolar amino acids with bulky side chains<br>Possible variants must be considered | |
| System T* | Prefers tryptophan to leucine, etc.<br>Human red cells | |
| System asc | Prefers threonine, serine, alanine, cysteine<br>Erythroid cells, but usually not in mature mammalian red cells. Analogy to systems ASC has not proved common origin | |
| **Systems for anionic amino acids** | | |
| $X_{A'}^- X_{G'}^-$ $X_{AG'}^- X_C^-$ | Identify systems preferring respectively aspartate, glutamate, neither, or cysteine and glutamate<br>Relation to neurotransmittor actions? | |
| **Systems for cationic amino acids** | | |
| $y^+$ | Lysine, arginine, ornithine<br>Two forms may appear together in gut. System specific to ornithine known in hepatic mitochondria | |

* With the two historical exceptions L and T, lower case letters are preferred for designating $Na^+$-independent transport systems.

recent assignment of amino acids to various carrier types. The absence or presence of a particular carrier type in a system depends on the demonstration of a saturable uptake by a suitable substrate of that class. With some classes of amino acid there are specific non-metabolized analogues which can be used to classify the types of carriers present, MeAIB is used to identify the 'A' system and BCH for the 'L'. However, there is a considerable degree of overlap, most amino acids having a weak affinity for other carrier types. This will result in them being transported or taken up to a small degree even though their 'true' carrier is absent. This can be seen, for example, for the uptake of glutamine and cysteine at the blood–brain barrier[16–18]. This weak 'non-specific' uptake can usually be identified by the fact that the defining substrate for that amino acid class, when present in excess, has no effect on the uptake of the labelled amino acid, whereas another 'unexpected' amino acid will cause a marked reduction in uptake. Table 3.2 shows a list of amino acids and whether their uptake has been demonstrated at the blood–brain barrier by a variety of methods. As can be seen from these tables the L and $y^+$ carrier types appear to be present at the luminal side of the blood–brain barrier whereas the A and Gly carriers are absent.

In order to improve the precision of the basic methods a number of *in situ* brain perfusions have been developed, as for example that of Gilboe and Betz, who used the indicator dilution technique to study the kinetics of leucine uptake in the isolated dog brain[19]. An alternative approach was that of Baños *et al.*[20,70], who used a quasi-steady state intravenous infusion technique in the rat, so that the concentrations of amino acids in the blood (in the presence of the animal's own amino acids) could be known and kept constant for a slightly longer time period. Bradbury used the mouse[21] and Smith the rat[22]. The experiments by the latter group used saline perfusions and periods of

**Table 3.2** The uptake of amino acids by the mammalian blood–brain barrier by various techniques. (+) indicates uptake, (s) indicates a small uptake, (–) indicates a non-significant uptake

| *Amino acid* | *SP* | *BUI* | *IVI* | *INS* |
|---|---|---|---|---|
| *Large neutral* | | | | |
| Cycloleucine | + | | | |
| Glutamine | – | s | | + |
| Histidine | + | + | + | + |
| Isoleucine | | + | + | + |
| Leucine | + | + | + | + |
| Methionine | | + | + | + |
| Phenylalanine | + | + | + | + |
| Threonine | | s | + | + |
| Tryptophan | +* | + | + | + |
| Tyrosine | + | + | + | + |
| Valine | + | + | + | + |
| *Small neutral* | | | | |
| Alanine | – | s | | s |
| Glycine | – | – | | |
| Proline | | – | | |
| Serine | – | s | | s |
| *Basic* | | | | |
| Arginine | s | + | + | |
| Lysine | – | + | + | |
| *Anionic* | | | | |
| Apartate | – | – | | |
| Glutamate | – | – | | |

SP = Single pass (dog)[10,11]; BUI = Brain uptake index (rat)[12,65]; IVI = Intravenous infusion technique (rat)[20,70]; INS = *In situ* brain perfusion (rat)[30–32]; * = Huet *et al.* (dog)[35]

only 60 seconds. Zlokovic, using the guinea-pig, chose an extended longer time course (20 min) to study the uptake substrates which cross the blood–brain barrier slowly[23]. The species that have been used vary in the anatomy of the circle of Willis, that in the dog, rat and mouse having a patent posterior communicating artery, whereas that in the guinea-pig and gerbil are less complete[24]. In those species where the circle is not fully patent, a true ipsilateral preparation can be obtained by keeping the perfusion pressure slightly above arterial, whereas in the others some crossing of substances to the contralateral hemisphere does occur. The advantage of a true single-sided preparation is that the actual concentration of amino acids in the perfusion fluid is known and not disturbed by the animal's homologous amino acids[23,32].

To quantitate the nature of the uptake processes for amino acids at the blood–brain barrier it is necessary to determine the transport parameters, $K_m$, $V_{max}$ and $K_d$. An absolute or 'true' value of these constants can be obtained by using non-plasma perfusion systems whereas 'apparent' values are derived if blood or plasma perfusate is used with their normal complement of amino acids. The transport parameters can also be influenced by the cerebral blood flow (CBF), pH, $PCO_2$ and anaesthesia which all must be considered in comparing the constants derived by the various methods.

With the paired tracer and the BUI methods, the actual concentration of an amino acid within the bolus depends on the mixing which will occur as it passes into the cerebral circulation[22,26], and this will also depend on the rate of CBF.

Some controversy exists with regard to the effect of anaesthesia on the derived kinetic constants from these methods. Sage and Duffy[27] found the BUI in the rat increased with anaesthesia whereas Hawkins *et al.*, using an autoradiographic method to investigate regional variation, found little or no difference between conscious and anaesthetized rats[28]. This was resolved by Pardridge[29], who pointed out that anaesthesia depresses CBF and so increases the extraction, but the actual amino acid influx was decreased in anaesthesia.

The actual value of $K_m$, $V_{max}$ and $K_d$ varies between the methods, firstly because in some preparations, such as that of Baños[20], there are natural plasma amino acids present and secondly, because of the nature of the technique. The lowest values reported for $K_m$ of amino acid uptake at the BBB are those of Smith *et al.* using the rapid brain perfusion method[22]. This method, it has been claimed, has the advantage over the others that the measurement of cerebrovascular permeability is 100 times more sensitive than the BUI, the rate of CBF is actually measured during the experiment (values are much greater than *in vivo*), there is negligible mixing or metabolic degradation in the infusate and that there is absolute control of the amino acid concentration as saline is used instead of plasma in the infusate. The technique which uses time periods up to 1 minute must require considerable agility to achieve measurements within subdivisions of a minute and although the animal is anaesthetized it does give low $K_m$ values. Pardridge, using the conscious rat[29], found values for the kinetic constants different from the Smith method and at present there is not a satisfactory explanation for these differences[30–32].

Although these differences in absolute values for amino acid kinetic constants are intriguing, what is most important is that the majority of the values of $K_m$ for amino acid uptake at the blood–brain barrier are close to the plasma concentrations found *in vivo*. This situation is markedly different from the other tissues of the body and will have a considerable effect on the availability of amino acids to the brain since, as has been pointed out by Fernstrom and Wurtman[33], the availability of an amino acid to the brain depends on the sum of the concentrations of the competing amino acids. If the values of $K_m$ are large the effect of competing amino acids is small, but if the $K_m$ is low and close or less than the plasma concentration, the entry into brain will be markedly sensitive to variation in the plasma concentration of any one of the competing amino acids. It is this fact which makes the brain vulnerable to hyperaminoacidaemias[29,34]. There are, however, some interesting species variations; measurements of $K_m$ in the conscious dog gave values which are much higher than in the conscious rat[29,35] so that the dog would appear to be less sensitive to variation in plasma amino acid levels than the other animals.

The movement of amino acids from blood into the brain across the blood–brain barrier is made up of two components. Firstly, there is a carrier-mediated step which is stereospecific and a facilitated process, with no sodium dependency[32,33]. The $K_m$ values for amino acids vary according to the method used and as to whether plasma amino acids are present, but, as can be seen in Table 3.3., the values are close to the plasma concentration for each amino acid. The $V_{max}$ values show much less variation than those for the $K_m$[30,31,33]. Secondly, there is also a non-saturable diffusion component $K_d$, which forms only a small contribution to the entry of amino acids at the normal physiological conditions, but if the plasma level rises, the contribution of this pathway, also increases and may become a significant part of the entry of amino acids from blood into brain[30–32]. The $K_m$ value of the cerebrovascular amino acid uptake system is much lower than in other organs, which indicates a high affinity, but a saturated system at normal plasma values[33,34]. Smith has calculated that the majority of the transport sites for neutral amino acids are occupied by PHE and LEU (>50%), the remainder being occupied by TRY, MET, ISOLEU, TYR, HIS, VAL, THR and GLU (2–7% each), small amino acids ALA, SER, CYS and ASP making a negligible contribution (<0.3%), with no uptake for GLY and PRO[31,32]. Competition between this group reduces the influx of amino acids and increases the value of $K_m$. This competition and saturation occurring at the BBB acts to prevent the daily fluctuations in plasma amino acids for altering the brain levels, so acting as an efficient homeostatic mechanism.

**Table 3.3** Kinetic constants $K_m$, $V_{max}$ and $K_d$ determined for the uptake of amino acids by the blood–brain barrier. 'Apparent' values refer to values measured in the presence of natural amino acids present either by mixing of the bolus or perfusate with blood during transit through the cerebral circulation. 'Real' values are obtained where no mixing is claimed to occur, so the 'true' concentration of amino acid is known

| | *Apparent* | | | *Real* | | |
|---|---|---|---|---|---|---|
| *Amino acids* | $K_m$ (mmol/l) | $V_{max}$ (nmol/min/g) | $K_d$ (μl/min/g) | $K_m$ (mmol/l) | $V_{max}$ (nmol/min/g) | $K$ (μl/min/g)d |
| Cycloleucine | [a]0.43 ± 0.19 | 27 ± 12 | 12 ± 6 | – | | |
| Histidine | [a]0.24 ± 0.002 | 29 ± 4 | 12 ± 3 | [m]0.10 ± 0.01 | 61 ± 2 | 3 ± 1 |
| | [b]0.16 ± 0.03 | 30 ± 7 | 33 ± 6 | | | |
| Isoleucine | [a]0.28 ± 0.06 | 42 ± 9 | NS | [q]0.06 ± 0.01 | 60 ± 4 | NS |
| | [b]0.15 ± 0.03 | 36 ± 9 | 31 ± 9 | | | |
| Leucine | [a]0.10 ± 0.01 | 22 ± 3 | 18 ± 4 | | | |
| | [b]0.09 ± 0.01 | 23 ± 4 | 45 ± 3 | [p]0.03 ± 0.01 | 59 ± 5 | 4 ± 1 |
| | [c]1.58 ± 0.28 | 323 ± 5 | | [s]0.13 ± 0.03 | 86 ± 7 | 13 ± 4 |
| | [d]1.50 ± 0.30 | 70 ± 3 | | | | |
| Methionine | [a]0.18 ± 0.08 | 29 ± 12 | 19 ± 11 | | | |
| | [b]0.08 ± 0.02 | 18 ± 5 | 45 ± 7 | [m]0.04 ± 0.02 | 25 ± 6 | 2 ± 1 |
| Phenylalanine | [a]0.11 ± 0.01 | 28 ± 7 | 14 ± 8 | | | |
| | [b]0.03 ± 0.01 | 14 ± 4 | 44 ± 14 | [n]0.011 ± 0.002 | 41 ± 4 | 8 ± 1 |
| Threonine | [e]0.73 | 37 | | | | |
| Tryptophan | [a]0.16 ± 0.01 | 30 ± 3 | NS | [m]0.22 ± 0.06 | 17 ± 4 | 4 ± 1 |
| | [b]0.05 ± 0.01 | 18 ± 5 | 41 ± 15 | [m]0.015 ± 0.003 | 55 ± 5 | 4 ± 1 |
| Tyrosine | [b]0.09 ± 0.02 | 25 ± 5 | 39 ± 7 | | | |
| | [a]0.15 ± 0.01 | 31 ± 8 | 16 ± 4 | [m]0.06 ± 0.01 | 97 ± 7 | NS |
| Valine | [a]0.51 ± 0.05 | 35 ± 3 | 3 ± 2 | | | |
| | [b]0.17 ± 0.07 | 14 ± 5 | 37 ± 7 | [m]0.21 ± 0.02 | 49 ± 3 | 5 ± 1 |
| Arginine | [a]0.09 ± 0.01 | 8 ± 1 | 4 ± 2 | | | |
| | [b]0.04 ± 0.02 | 5 ± 3 | 26 ± 2 | – | – | – |
| | [f]0.10 ± 0.04 | 10 ± 1 | – | | | |
| Lysine | [a]0.11 ± 0.02 | 8 ± 1 | NS | – | – | – |
| Ornithine | [a]0.22 ± 0.04 | 11 ± 2 | NS | – | – | – |

NS = not significant different from zero; [a]Pardridge (1983) rat[66]; [b]Miller *et al.* (1985) rat[29]; [c]Betz *et al.* (1975) dog[19]; [d]Daniel *et al.* (1977) rat[62]; [e]Pardridge and Oldendorf (1975) rat[64]; [f]Oldendorf *et al.* (1983) rat[16]; [m]Smith *et al.* (1987) rat[32]; [n]Momma *et al.* (1987) rat[31]; [p]Smith *et al.* (1985) rat[30]; [q]Takasato *et al.* (1985) rat[22]; [s]Pratt (1985) rat[67]

## THE BRAIN SIDE OF THE CEREBRAL CAPILLARY ENDOTHELIUM

### Isolated capillaries

It is difficult to gain access to the 'hidden' brain side of the cerebral capillaries. Two methods have been used, firstly separation of cerebral capillaries from brain homogenates and measurement of the uptake of substances into the capillary tubular fragments[25], and secondly, by culture of the cerebral endothelium as a sheet on a support grid[36,37], The pioneers in the use of isolated capillary tubules were Betz and Goldstein[36], who demonstrated that this tissue was able to take up both 'L' and 'A' amino acids or their analogues. Leucine was used as a typical 'L' amino acid and MeAIB for the 'A' system. The uptake of MeAIB was shown to be sodium dependent, which is typical of the 'A' system, whereas the uptake of leucine was sodium-independent. The MeAIB uptake was inhibited by proline and methionine, neither of which interfered with the uptake of leucine. Some overlap in uptake was also shown, since leucine and phenylalanine caused some inhibition of the 'A' system uptake[36]. Since no 'A' system has been found *in vivo* on the blood side of the BBB, there can be little doubt that the 'A' carrier is present on the brain side of the cerebral endothelium. Further confirmation of this finding was that collagenase treatment removed only the 'A' uptake and left the 'L' system intact[38]. It has been argued that the lengths of broken capillary tubule are open at each end, so the medium gains access to both sides of the capillary wall and the uptake will reflect carrier mediated uptake on both 'faces' of the cell[39]. The preparations also contain arterioles, glia and some neural tissue. A possibility could be that the collagenase treatment removes the astrocytic end-feet which envelope the capillaries and the 'A' system uptake is on this glial tissue. This has been refuted by Cardelli-Cangiano *et al.* who argue that the collagenase removes $\gamma$-glutamyl transpeptidase and it is the loss of this enzyme which results in the failure of treated capillaries to take up glutamine[38]. It is of interest that the uptake of BCH, an 'L' system analogue, is enhanced by collagenase treatment.

In conclusion it would appear that there are 'A' system carriers on the brain side of the cerebral capillaries and these carriers probably function to remove amino acids with transmitter activity from the brain ISF and, in addition, this process may be coupled to the entry of large neutral amino acids moving from blood into brain[40,41].

### Cultured brain capillary endothelial monolayers

An alternative approach has been to culture the cerebral capillary endothelium as a monolayer in various types of support grids. These discs can then be placed between the two halves of a modified Ussing chamber, so separating the blood side of the endothelium from the brain side[42–44]. This technique can be used to study transendothelial transport instead of cellular uptake, and the factors which influence the net direction of transport of various molecules.

As with the isolated capillary tubular fragments, the 'L' and 'A' systems appear to be present, but the 'A' system transports symmetrically in both directions. If, however, $C_6$ glioma cells (astrocytes) are grown on the underside of the support grid or the cells are exposed to glial conditioned medium, the monolayer becomes polarized and there is a net transport from the glial side towards the 'blood' side of the endothelium[45,46]. This finding does not support the arguments put forward by Cardelli-Cangio[38] for the effect of collagenase on isolated brain capillary tubular fragments. The kinetics of leucine transport across these monolayers appear to be similar to those determined *in vivo*[41,43], and specific for large neutral amino acids. Protein synthesis inhibition caused a marked increase in the transendothelial flux of leucine, since more was available for transport and not lost into cellular protein synthesis[43]. However, although this approach is of considerable interest, the endothelial sheet is still more than a hundred times more permeable to sucrose than the *in vivo* BBB[43]. Similar problems have been encountered with other epithelial monolayers cultured *in vitro* and it may be the lack of hormones or the excess of nutrients and oxygen on the basal side of the sheet which disturbs the formation of true tight junctions and prevents the development of full polarity of the cell layer. In support of this hypothesis, cells grown on the base of tissue culture dishes show normal expression of proteins where as those grown on grids do not[47].

## Exchange of amino acids between the brain ISF, neurons and glia

The entry of amino acids across the BBB into the brain ISF is a sodium-independent facilitated process. The efflux of amino acids from brain ISF to blood may be sodium-dependent for the 'A' amino acids and also uphill (see previous section).

The neurons and glia can themselves take up amino acids from the brain ISF, so can exert some control on the composition of amino acids in the ISF. Studies by McIlwain, Bachelard and others, using brain slices, have demonstrated that the uptake of amino acids is sodium-dependent and influenced by insulin[48,49]. More recently, Sershen and Lajtha have examined the various classes of carrier types and have identified some ten different carriers. Five were the A, L and ASC, anionic and cationic systems as described by Christensen and the remainder were specific for glycine, GABA, taurine and lysine[50]. This complexity is necessary to control these amino acids which also have a neurotransmitter role.

## Brain ISF/CSF relations

The concentration of amino acids in CSF is 25% or less of that found in plasma[24]. If it is assumed that the brain ISF has a concentration of amino acids similar to that in plasma, there will be a concentration gradient between the two fluids. Since the ependymal lining of the ventricles is highly permeable to all molecules, it can be seen that the CSF will act as a 'sink' into which the brain ISF amino acids can diffuse[24]. Cserr has demonstrated a bulk flow of brain ISF into CSF[51], so this type of system, with a restricted entry, active uptake by neurons and a bulk flow drainage into CSF, could act as an efficient homeostatic mechanism for brain ISF amino acid control.

## CSF and the choroid plexus

The low level of amino acids in CSF with respect to plasma could be the result of three processes. Firstly, the entry of amino acids across the choroid plexus into the nascent CSF could be restricted, so the concentration in this newly formed fluid would be low. Secondly, the choroid plexuses could actively remove amino acids from CSF to blood, and thirdly, the neurons and glia may take up amino acids as the CSF flows along its normal circulation pathways.

Taking the last probability first, studies with glucose by Davson and Bito[52] failed to show any difference in concentration between CSF sampled from the lateral ventricles, cisterna magna or subarachnoid space, so this possibility would seem unlikely. With regard to the possible uptake of amino acids by the choroid plexus from CSF, studies *in vivo* using the technique of ventriculo-cisternal perfusion (Vc)[53] have shown that the concentration of amino acids leaving the cisterna magna is much less than that entering the ventricles. In addition, the incubated non-perfused choroid plexuses *in vitro*, show considerable ability in concentrating amino acids which are sodium dependent and the tissue/medium (T/M) ratio achieved after incubation is much greater than 1[54]. Studies by Davson and his colleagues using Vc perfusions have also shown that the uptake of amino acids is carrier mediated and will work 'uphill'[55]. These findings have lead to the postulate that the low level of amino acids in CSF is a consequence of the uptake of amino acids out of CSF by the choroid plexuses. Exactly the same situation had previously occurred with sugars and it was not until the isolated blood perfused choroid plexus of the sheep was used that the situation was elucidated[56]. With this preparation and by the use of $^{3}H$- and $^{14}C$-labelled sugars, it is possible to measure the net flux of sugars across the choroid plexus in the same preparation and in these experiments it was found that the net direction of transport was from blood to CSF and not in the opposite direction[57].

There is no doubt that there is an active uptake of sugars by the choroid plexus out of CSF which is sodium-dependent but the $K_m$ of these processes indicates that it is saturated at the normal values found in CSF and the $V_{max}$

is relatively low. These kinetic data indicate that the function of this uptake is probably part of the normal cellular processes found in most cells and is insufficient in its capacity to be responsible for the lowering of the CSF glucose concentration. From a functional point of view it would seem inefficient to 'filter' amino acids at the same concentration as in plasma into the CSF then recover them at high energy costs within the same tissue.

Using the isolated perfused sheep choroid plexus, amino acid fluxes have been measured both from CSF to Ringer and from Ringer to CSF, using the steady-state method with the same concentration of amino acid in both fluid compartments. At amino acid concentrations less than those normally found in mammalian plasma (10 $\mu$mol/l), fluxes were measured in both directions, with a net amino acid flux from the Ringer into CSF, similar to that for sugars[57]. However, as the concentration was raised beyond that found in plasma, efflux from the CSF increased whilst the flux from Ringer remained at a low level, suggesting saturation of specific transport systems. Under these circumstances the net flux reversed direction with amino acid movement from the CSF to Ringer being the most prominent. The concentration at which reversal occurred varied for each amino acid tested, but in general it was around that normally found in plasma, or slightly below, especially for the putative neurotransmitter glycine where a net CSF to Ringer flux was measured at a concentration of 100 $\mu$mol/l, approximately one third of that found in plasma[68,69,70,72].

Preliminary kinetic estimations of CSF to Ringer flux suggest Michaelis–Menten $K_m$ values between 0.3 mmol/l (glycine) and 2.1 mmol/l (L-phenylalanine) values of a similar magnitude to those measured during ventriculo-cisternal perfusion in the cat[58]. The comparable values for Ringer to CSF flux were estimated as 0.02 mmol/l for glycine and 0.003 mmol/l for L-phenylalanine, suggesting much higher affinity carrier systems for these amino acids at the basolateral face (blood side) of the choroid plexus. The $V_{max}$ for CSF to Ringer flux was around one order of magnitude greater than that measured for Ringer to CSF flux[69,72].

These results lend support both to the concept of restricted entry of amino acids from the blood and to clearance of amino acids from CSF, a combination of the two providing the necessary homeostasis of levels in CSF regardless of the plasma amino acid concentrations.

To study the uptake processes on the blood-side of the choroid plexus, the indicator dilution technique has been used, so that the receptor types of a wide range of amino acids could be identified. As can be seen in Figure 3.1, there are carriers for the large neutral amino acids (LNAA), such as leucine, phenylalanine, tryptophan, etc., the same as at the BBB. Marked differences, however, were found with the cationic and anionic amino acids which had uptake values of a similar order to that of the LNAA. In addition, there are specific carriers for glycine, proline and glutamine. The uptake of alanine is low, and since MeAIB is not taken up, it would appear that the 'A' carrier is absent from this side of the choroid plexus which is similar to that of the BBB.

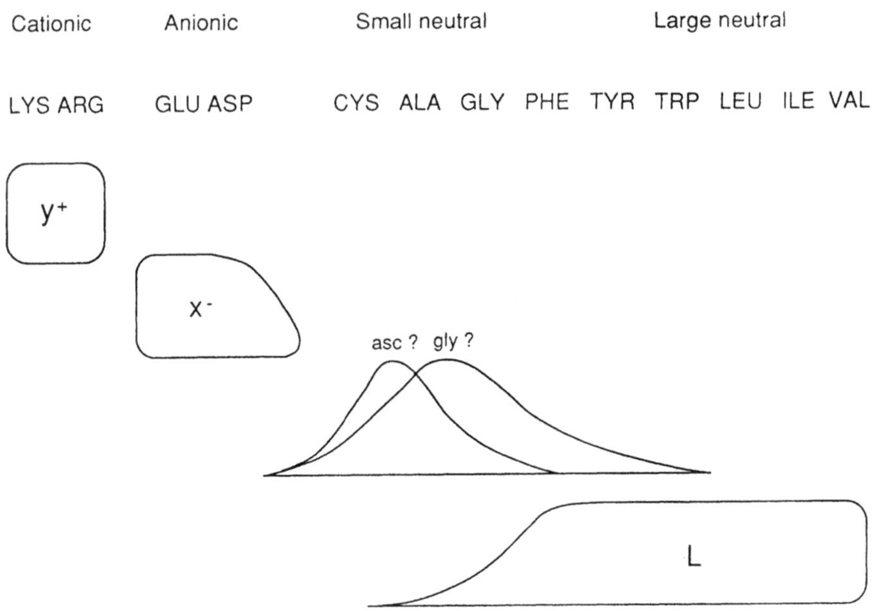


**Figure 3.1** Schematic representation of amino acid carrier types at the basolateral face of the isolated perfused sheep choroid plexus

However, unlike the BBB where there is an 'A' carrier system on the brain side, this is absent from the CSF side of the choroid plexus as shown by Davson's Vc studies[55].

In conclusion it would appear that the amino acids, like the sugars, are restricted at the point of entry across the choroid plexus but with some amino acids with a neurotransmitter role the choroid plexuses may have a 'scavenger' function similar to the BBB.

## Arachnoid/CSF exchanges

The arachnoid membranes have been shown to be active in the uptake of amino acids. The clearance of amino acids from ventricle to subarachnoid space is much greater in than the Vc perfusions[59]. Similarly, the isolated frog choroid plexus shows no net direction for the movement of amino acids, whereas the isolated arachnoid membrane from the same species has a marked net uptake of glycine from CSF to blood[60,61].

## References

1. Bradbury, M.W.B. (1979). *The Concept of the Blood–Brain Barrier*. (Wiley, Chichester)

2. Kamin, H. and Handler, P. (1951). The metabolism of parenterally administered amino acids II. Urea synthesis. *J. Biol. Chem.*, **188**, 193–205
3. Chirigos, M.A., Greengard, P. and Udenfriend, S. (1960). Uptake of tyrosine by rat brain *in vivo*. *J. Biol. Chem.*, **235**, 2075–2079
4. Udenfriend, S. (1961). Phenylketonuria. *Am. J. Clin. Nutrit.*, **9**, 691–694
5. Guroff, G. and Udenfriend, S. (1962). Studies on aromatic amino acid uptake by rat brain *in vivo*. *J. Biol. Chem.*, **237**, 803–806
6. Lajtha, A. and Toth, J. (1961). Uptake and transport of amino acids by the brain. *J. Neurochem.*, **8**, 216–225
7. Chinard, F.P., Vosburgh, G.J. and Enns, T. (1955). Transcapillary exchange of water and of other substances in certain organs of the dog. *Am. J. Physiol.*, **183**, 221–234
8. Crone, C. (1961). Gu diffusionen of nogle organisk non-elektrolyter fra clod til hjeznevaeu. (Munksgaard, Copenhagen)
9. Yudilevich, D.L. and De Rose, N. (1971). Blood–brain transfer of glucose and other molecules by rapid indicator dilution. *Am. J. Physiol.*, **220**, 841–846
10. Yudelivich, D.L., De Rose, N. and Sepulveda, F.V. (1972). Facilitated transport of amino acids through the blood–brain barrier of the dog studied in a single capillary circulation. *Brain Res*, **44**, 569–578
11. Yudilevich, D.KL. and Sepulveda, F.V. (1976). The specificity of amino acid and sugar carriers in the capillaries of the dog brain studied *in vivo* by rapid indicator dilution. In *Transport Phenomena in the Nervous System*, eds., G. Levi, L. Battistin and A. Lajtha, A (Plenum, New York), pp. 77–85
12. Oldendorf, W.H. and Szabo, J. (1976). Amino acid assignment to one of three blood–brain barrier amino acid carriers. *Am. J. Physiol.*, **230**, 94–98
13. Oldendorf, W.H. (1981). Clearance of radio labelled substances by brain after arterial injection using a diffusible internal standard. In eds., N. Marlis and R. Rodright *Research Methods in Neurobiochemistry*, Vol. 5, (Plenum, New York), pp. 91–112
14. Christensen, H.N. (1969). Some special kinetic problems of transport. *Adv. Enzymol.*, **32**, 1–31
15. Christensen, H.N. and Kilberg, M.S. (1987). Amino acid transport across the plasma membrane: role of regulation in interorgan flow. In: *Amino Acid Transport in Animal Cells*. eds. D.L. Yudilevich and C.A.R. Boyd (MUP, Manchester)
16. Oldendorf, W.H., Crane, P.D., Brown, L.D., Wade, L.A. and Diamond, J.N. (1983). Blood brain barrier transport of basic amino acids is selectivity inhibited by low pH. *J. Neurochem.*, **40**, 797–800
17. Pardridge, W.M. and Mietus, L.J. (1982). Kinetics of neutral amino acid transport through the blood–brain barrier of the newborn rabbit. *J. Neurochem.*, **38**, 955–962
18. Wade, L.A. and Brady, H.M. (1981). Cysteine and cystine transport at the blood–brain barrier. *J. Neurochem.*, **37**, 730–734
19. Betz, A.L., Gilboe, D.D. and Drewes, L.R. (1975). Kinetics of unidirectional leucine transport into brain: effects of isoleucine, valine and anoxia. *Am. J. Physiol.*, **228**, 895–902
20. Baños, G., Daniel, P.M. and Pratt, O.E. (1974). Saturation of shared mechanism which transports L-arginine and L-lysine into the brain of the living rat. *J. Physiol.*, **236**, 29–41
21. Bradbury, M.W., Deane, R. and Rosenberg, G. (1984). Regional blood flow, EEG and electrolytes in mouse brain perfused with perfluorochemical, FC43. *J. Physiol.*, **355**, 31
22. Takasato, Y., Rapoport, S.I. and Smith, Q.R. (1984). An *in situ* brain perfusion technique to study cerebrovascular transport in the rat. *Am. J. Physiol.*, **247**, H484–H493
23. Zlokovic, B.V., Begley, D.J., Duricic, B.M. and Mitrovic, D.M. (1986). Measurement of solute transport across the blood–brain barrier in the perfused guinea-pig brain, method and application to *N*-methyl-aminoisobutyric acid. *J. Neurochem.*, **46**, 1444–1451
24. Davson, H., Welch, K. and Segal, M.B. (1987). *The Physiology and Pathology of the Cerebrospinal Fluid*. (Churchill Livingstone, Edinburgh)
25. Joó, F. (1985). The blood–brain barrier *in vitro*: ten years of research on microvessels isolated from the brain. *Neurochem. Int.*, **7**, 1–25

26. Pardridge, W.M., Landaw, E.M., Miller, L.P., Brown, L.D. and Oldendorf, W.H. (1985). Carotid artery injection technique: bounds for bolus mixing by plasma and by brain. *J. Cereb. Blood Flow Metab.*, **5**, 576–583
27. Sage, J.I. and Duffy, T.E. (1979). Pentobarbital anaesthesia: influence on amino acid transport across the blood–brain barrier. *J. Neurochem.*, **33**, 963–965
28. Hawkins, R.A., Mans, A.M. and Bieuyck, J.F. (1982). Amino acid supply to individual cerebral structures in awake and anaesthetized rats. *Am. J. Physiol.*, **242**, E1–E11
29. Miller, L.P., Pardridge, W.M., Brown, L.D. and Oldendorf, W.H. (1985). Kinetic constants for blood–brain barrier amino acid transport in conscious rats. *J. Neurochem.*, **45**, 1427–1432
30. Smith, Q.R., Takasato, Y., Sweeney, D.J. and Rapoport, S.I. (1985). Regional cerebrovascular transport of leucine as measured by the *in situ* brain perfusion technique. *J. Cerebr. Blood Flow Metab.*, **5**, 300–311
31. Momma, S., Aoyagi, M., Rapoport, S.I. and Smith, Q.R. (1987). Phenylalanine transport across the blood–brain barrier as studied with the *in situ* brain perfusion technique. *J. Neurochem.*, **48**, 1291–1300
32. Smith, Q.R., Momma, S., Aoyagi, M. and Rapoport, S.I. (1987). Kinetics of neutral amino acid transport across the blood–brain barrier. *J. Neurochem.*, **49**, 1651–1658
33. Fernstrom, J.D. and Wurtman, R.J. (1972). Brain serotonin content: physiological regulation by plasma neutral amino acids. *Science*, **178**, 414–416
34. Pardridge, W.M. and Choi, T.B. (1986). Neutral amino acid transport at the human blood–brain barrier. *Fed. Proc.*, **45**, 2073–2078
35. Huet, P.-M., Pomier-Layrargues, G., Duguay, L. and Du Souich, P. (1981). Blood–brain transport of tryptophan and phenylalanine; effect of porto-caval shunt in dogs. *Am. J. Physiol.*, **241**, G163–169
36. Betz, A.L. and Goldstein, G.W. (1978). Polarity of the blood– brain barrier: neutral amino acid transport into isolated brain capillaries. *Science*, **202**, 225–226
37. Audus, K.L. and Borchardt, R.T. (1986). Characteristics of an *in vitro* blood–brain barrier model system for studying drug transport and metabolism. *Pharm. Res.*, **3**, 81–87
38. Cardelli-Cangiano, P., Fiori, A., Cangiano, C., Barberini, F., Allegra, P., Peresempio, V. and Strom, R. (1987). Isolated brain microvessels as *in vitro* equivalents of the blood–brain barrier: selective removal by collagenase of the "A"-system of neutral amino acid transport. *J. Neurochem.*, **49**, 1667–1675
39. Bradbury, M.W. (1985). Critique. In: The blood–brain barrier *in vitro*: ten years of research in microvessels isolated from the brain. *Neurochem. Int.*, **7**, 27–28
40. Cangiano, C., Cardelli-Cangiano, P., James, J.H., Rossi-Fanelli, F., Patrizi, M.A., Brackett, K.A., Strom, R. and Fischer, J.E. (1983). Brain microvessels take up large neutral amino acids in exchange for glutamine: cooperative role of $Na^+$-dependent and $Na^+$-independent systems. *J. Biol. Chem.*, **258**, 8949–8954
41. Betz, A.L. and Goldstein, G.W. (1986). Specialized properties and solute transport in brain capillaries. *Ann. Rev. Physiol.*, **48**, 241–250
42. Ussing, H.H. and Zerahn, K. (1951). Active transport of sodium as the source of electric current in the short-circuited isolated frog skin. *Acta Physiol. Scand.*, **23**, 110–127
43. Audus, K.L. and Borchardt, R.T. (1986). Characteristics of the large neutral amino acid transport system of bovine brain microvessel endothelial cell monolayers. *J. Neurochem.*, **47**, 484–488
44. Bowman, P.D., Ennes, S.R., Rarey, K.E., Betz, A.L. and Goldstein, G.W. (1983). Brain microvessel endothelial cells in tissue culture: a model for study of blood–brain barrier permeability. *Ann. Neurol.*, **14**, 396–402
45. Cancilla, P.A. and De Bault, L.E. (1983). Neutral amino acid transport properties of cerebral endothelial cells *in vitro*. *J. Neuropath. Exp. Neurol.*, **42**, 191–189
46. Beck, D.W., Vintners, H.V., Hart, M.N. and Cancilla, P.A. (1984). Glial cells influence polarity of the blood–brain barrier. *J. Neuropath. Exp. Neurol.*, **43**, 219–224

47. Schreiber, G. Department of Biochemistry, Melbourne University, Australia. Personal communication
48. McIlwain, H. and Bachelard, H.S. (1985). *Biochemistry of the Central Nervous System*. (Churchill Livingstone, Edinburgh), pp. 154–201
49. McIlwain, H. (1975). *Practical Neurochemistry*. (Churchill Livingstone, Edinburgh)
50. Sershen, H. and Lajtha, A. (1979). Inhibition pattern by analysis indicates the presence of ten or more transport systems for amino acids in brain cells. *J. Neurochem.*, **32**, 719–726
51. Cserr, H.F., Cooper, D.N. and Milhorat, T.H. (1977). Flow of cerebral interstitial fluid as indicated by the removal of extracellular markers from rat caudate nucleus. *Exp. Eye Res.*, **25** (suppl)., 461–473
52. Bito, L.Z. and Davson, H. (1966). Local variations in cerebrospinal fluid composition and its relationship to the composition of the extracellular fluid of the cortex. *Exp. Neurol.*, **14**, 264–280
53. Snodgrass, S.R., Cutler, R.W.P., Kang, E.S. and Lorenzo, A.V. (1969). Transport of neutral amino acids from the feline CSF. *Am. J. Physiol.*, **217**, 974–980
54. Lorenzo, A.V. (1977). Factors governing the composition of the cerebrospinal fluid. *Exp. Eye Res.*, **25** (suppl)., 205–228
55. Davson, H., Hollingsworth, J.G., Carey, M.B. and Fenstermacher, J.D. (1982). Ventriculo-cisternal perfusion of twelve amino acids in the rabbit. *J. Neurobiol.*, **13**, 293–318
56. Segal, M.B. and Pollay, M. (1977). The secretion of cerebrospinal fluid. *Exp. Eye Res.*, **25** (suppl)., 127–148
57. Deane, R. and Segal, M.B. (1985). The transport of sugars across the perfused choroid plexus of the sheep. *J. Physiol.*, **362**, 245–260
58. Lorenzo, A.V. (1974). Amino acid transport mechanisms of the cerebrospinal fluid. *Fed. Proc.*, **33**, 2079–2085
59. Lorenzo, A.V. and Snodgrass, S.R. (1972). Leucine transport from the ventricles and the cranial subarachnoid space in the cat. *J. Neurochem.*, **19**, 1287–1298
60. Wright, E.M. (1972). Accumulation and transport of amino acids by the frog choroid plexus. *Brain Res.*, **44**, 207–219
61. Wright, E.M. (1974). Active transport of glycine across the frog arachnoid membrane. *Brain Res.*, 76, 354–358
62. Daniel, P.M., Pratt, O.E. and Wilson, P.A. (1977). The transport of L-leucine into brain of the rat *in vivo*: saturable and non saturable components of influx. *Proc. R. Soc. B.*, **196**, 333–346
63. Yudilevich, D.L., Sepulveda, F.V., Bustamente, J.C. and Mann, G.E. (1979). A comparison of amino acid transport and ouabain binding in the brain endothelium studied *in vivo* by rapid paired tracer dilution. *J. Neurol. Trans.*, **10**, 15–27
64. Pardridge, W.M. and Oldendorf, W.H. (1975). Kinetics of blood–brain barrier transport of amino acids. *Biochim. Biophys. Acta*, **401**, 128–136
65. Oldendorf, W.H. (1971). The brain uptake of radiolabeled amino acids, amines and hexoses after intra-arterial injection. *Am. J. Physiol.*, **221**, 1629–1639
66. Pardridge, W.M. (1983). Brain metabolism: a perspective from the blood–barrier. *Physiol. Rev.*, **63**, 1481–1583
67. Pratt, O.E. (1985). Continuous injection method for the measurement of flux across the blood–brain barrier. The steady state, initial rate method. In: *Research Methods in Neurochemistry*, eds., Marks, N. and Rodnight, R. (Plenum Press, New York), Vol. 6, pp. 117–150
68. Preston, J.E., Segal, M.B., Walley, G.J. and Zlokovic, B.V. (1989). Neutral amino acid uptake by the isolated perfused sheep charoid plexus. *J. Physiol.*, **408**, 31–43
69. Segal, M.B., Preston, J.E., Collis, C.S. and Zlokovic, B.V. (1990). Kinetics and Na-independence of amino acid uptake by the blood-side of the perfused choroid plexus of the sheep. *Am. J. Physiol.* (in press)

70. Preston, J.E. and Segal, M.B. (1990). The uptake of anionic and cationic amino acids by the isolated perfused sheep choroid plexus. (in press)
71. Baños, G., Daniel, P.M., Moorhouse, S.R. and Pratt, O.E. (1973). The influx of amino acids into the brain of the rat *in vivo*: the essential compared with some non-essential. *Proc. R. Soc. Lond. B.*, **183**, 59–70
72. Preston, J.E. Ph.D. Thesis, London University, 1989

# 4 The role of peptides in the brain

## CENTRAL NERVOUS SYSTEM EFFECTS OF PEPTIDES

In the last 10 years there has been an enormous expansion of research on the effect of peptides on the central nervous system. The field has advanced rapidly, which has led to the development of the concept of *multiple peptide actions* in the central nervous system, under both physiological and pathological conditions. A number of recent experimental and clinical studies have provided strong evidence that the function by which most peptides were originally discovered and characterized is not their usual action, and is often not their most important action. However, from these early studies we have inherited most of the current peptide nomenclature, which in many cases was based on the original proposed function of the peptide.

The role of peptides in the central nervous system is complex and multiple, underlying the basic physiological phenomena of the brain. This involves: (a) neurotransmission and/or neuromodulation, (b) regulation of the neuroendocrine axis, (c) regulation of cerebral blood flow, (d) regulation of cerebrospinal fluid secretion, (e) mediation of the integrity of the blood–brain barrier, (f) modulation of the blood–brain barrier permeability to nutrients, and (g) regulation of water and electrolyte content of the brain. In the first section of this chapter we will focus primarily on the *behavioural* effects of peptides, as well as on their effects on the *higher integrative functions* of the brain and on the *vegetative* functions of the central nervous system. In the following sections we will discuss the other functions of peptides in the brain.

A glossary of the most commonly used brain peptide abbreviations is given in Table 4.1.

**Table 4.1** Glossary of the most commonly used brain peptide abbreviations.

| *Peptide* | *Abbreviation* |
|---|---|
| Adrenocorticotrophin | ACTH |
| Atrial natriuretic factor | ANF |
| $\beta$-endorphin | $\beta$-E or Endorphin (beta) |
| $\beta$-lipoprotein | $\beta$-LPH |
| Bombesin | BBS |
| Calcitonin | CT |
| Calcitonin gene-related peptide | CGRP |
| Cholecystokinin | CCK |
| Corticotrophin releasing factor | CRF |
| Delta sleep inducing peptide | DSIP |
| Eledoisin | Eled |
| Enkephalin leucine | Enk(Leu) |
| Enkephalin methionine | Enk(Met) |
| Epidermal growth factor | EGF |
| Fibroblast growth factor | FGF |
| Follicule-stimulating hormone | FSH |
| Gastrin releasing peptide | GRP |
| Gonadotrophin releasing hormone | GnRH |
| Growth hormone | GH |
| Growth hormone releasing factor | GRF |
| Insulin like growth factor | IGF |
| Luteinizing hormone | LH |
| Luteinizing hormone releasing hormone | LHRH |
| Melanocyte stimulating hormone ($\alpha$) | $\alpha$-MSH |
| Melanotrophin (MSH) release-inhibiting factor-1 | MIF-1 |
| Neural growth factor | NGF |
| Neuropeptide Y | NPY |
| Neurotensin | NT |
| Pancreatic polypeptide | PP |
| Parathyroid hormone | PTH |
| Peptide YY | PYY |
| Prolactin | PRL |
| Pro-opiomelanocortin | POMC |
| Somatostatin | SRIF |
| Substance K | SK |
| Substance P | SP |
| Thyroid stimulating hormone | TSH |
| Thyrotrophin releasing hormone | TRH |
| Tyr-Pro-Leu-Gly-amide | Tyr-MIF-1 |
| Vasoactive intestinal peptide | VIP |
| Vasopressin arginine | VP(Arg) |
| Vasopressin lysine | VP(Lys) |
| Vasotocin arginine | VT(Arg) |

## The route of peptide administration

A further glossary of the abbreviations used for the route of administration of peptides employed for studying their central actions is given in Table 4.2.

The most frequently studied naturally occurring brain peptides are given in Table 4.3, which also shows the peptide, the route of administration, the species, the observed central nervous effects and the direction of change, if any. This tabular synopsis provides a selected current review with respect to the multiple central nervous effects of peptides and is based on the results reported in articles which were published during the period from 1980 to 1985 in *Peptides*. This period and this journal have been chosen as a representative sample of the topic, with an aim to keep this review to a manageable size, but at the same time to emphasize the variety of central actions of the peptides rather than to enumerate most of them. In different sections of this book many other well-known peptide effects on the brain are described. From Table 4.3 it can be seen that the effect on the brain of a peptide is markedly influenced by the route of administration. Thus, for example, the analgesic effects of naturally occurring endogenous opiates in humans are more pronounced after an i.c.v., than after than i.v. administration[2]. Similarly, i.c.v. injection of endorphin ($\alpha$, $\beta$, $\gamma$) in the chick reduces distress vocalization, while i.p. injection is without effect[3]. On the other hand, i.p. injection of Enk in mice produces a marked analgesic effect, which supports the view that enkephalins may affect the brain after systemic administration[4].

**Table 4.2** Glossary of the routes of peptide administration used for studying their central actions

| *Route of administration* | *Abbreviations* |
|---|---|
| **Peripheral (systemic)** | |
| Intra-arterial | i.a. |
| Intra-muscular | i.m. |
| Intra-nasal | i.n. |
| Intra-peritoneal | i.p. |
| Intra-venous | i.v. |
| Sub-cutaneous | s.c. |
| **Local** | |
| Microinjections by iontophoretic application into specific brain region | cns. |
| Intracerebroventricular | icv. |
| Intracisternal | icis. |
| Intrathecal (spinal) | it. |
| Third ventricle | 3V |
| Fourth ventricle | 4V |
| Ventriculocisternal | vc. |

Some of the peptide effects can be obtained only by using a certain route of administration. Thus, for instance, an increased rumination in the sheep can be obtained only after i.c.v. injection of gastrin 1-17, while the systemic administration of the same peptide was ineffective[5]. On the other hand, both i.c.v. and i.v. injections of different gastrin analogues, gastrin-penta and -tetra in the sheep, caused an increased rumination[5].

An important characteristic for most peptides is that they can have central actions even after systemic administration, providing they are given in a sufficiently high pharmacological dose[1,6]. This implies that they can reach the brain by way of the *general circulation* assuming of course that the primary site of action is not localized in the periphery. Thus, the metabolic, transport and receptor events occurring during an interaction of the circulating peptide with either the blood–brain barrier, the blood–CSF barrier or circumventricular organs must be intimately involved in the mechanism by which peptides exert their action on the central nervous system.

## Behavioural effects

As can be seen in Table 4.3 and Table 4.4, peptides can cause marked changes in the behaviour of man and animals. Behaviour patterns such as drinking, feeding, alcohol intake and satiety can be modified and levels of aggression, grooming, locomotor activity, mating and sleep can also be altered by the administration of the appropriate peptides. More subtle changes can be recorded as changes in the electroencephalogram, in stereotype and in the depression of induced seizures after the administration of some peptides.

As relatively large doses of peptides were used in some of these studies, the reader is cautioned against drawing the conclusion that any or all of these peptides are involved in the normal physiological regulation of the above mentioned and tabulated behavioural functions.

## Peptides and higher integrative brain functions

It has been more than 20 years since de Wied and his co-workers suggested that vasopressin may be involved in memory. This hypothesis was derived from the observations that after the administration of vasopressin to rats, which had their posterior pituitaries removed, or to Battleboro rats which are naturally deficient in vasopressin, the rate of extinction of the avoidance response was restored to normal. On the basis that a delayed rate of extinction of a response reflects an enhanced memory it has been suggested that vasopressin-modulated processes may underlie the mechanism(s) of memory consolidation.

**Table 4.3** Multiple central nervous effects of the most frequently studied naturally occurring brain peptides modified according to Zadina *et al.*, (1986)[1].

| *Peptide* | *Route* | *Effect* | *Species* | *Change* | *Ref.* |
|---|---|---|---|---|---|
| ACTH 1-24 | icv | Grooming | Rat | ↑ | 7 |
| ACTH 1-39 | i.p. | Motor unit activity | | ↓ | 8 |
| | s.c. | Locomotor activity (open field) | | ↑ | 9 |
| ACTH | cns. (preopotic area) | | | ↓ | 10 |
| Angiotensin II | icis. | Sympathetic outflow $\beta$-E induced | | ↓ | 11 |
| | icv. | Drinking | | ↓ | 12 |
| | i.v. | Blood pressure | | ↓ | 13 |
| | | Glucose utilization (subfornical organ) | | ↓ | 13 |
| BBS | icis. | Dopamine turnover (striatum, olfactory | | ↓ | 14 |
| | | tubercle, hypothalamus) Gastric secretion | | ↑ | 15 |
| | icv. | Electroencephalogram | Rabbit | Complex biphasic | 16 |
| CGRP | cns. | Neuronal activity | Rat | ↑ | 22 |
| | icv.,it. | Gastrointestinal motility, fasted pattern | | Restored in rat | 40 |
| CRF | cns. | Neuronal activity | | ↑ | 41 |
| | icv. | Arousal | Monkey | ↓ | 42 |
| | | Locomotor activity | Rat | ↓ | 44 |
| | i.v. | Blood pressure, heart rate | | ↑ | 44 |
| DSIP | i.p. | Locomotor activity | Mouse | ↑ | 45 |
| | s.c. | Memory, short-term delayed response | Monkey | ↑ | 46 |
| Dynorphin 1-13 | icv. | Analgesia, morphine-induced (brain) | Rat | ↑ | 47 |
| Dynorphin | | Temperature | | ↑ | 48 |
| Eledoisin | icv. | Drinking | Pigeon | ↓ | 49 |
| $\alpha$-E | | Distress vocalization | Chick | ↑ | 50 |
| | i.p. | | | 0 | 50 |
| $\beta$-E | cns. (preoptic area) | Metabolic rate | Rat | ↓ | 10 |
| | | Temperature | | ↓ | 10 |
| | icis. | Gastric secretions | | ↑ | 15 |
| | icv. | Distress vocalization | Chick | ↑ | 50 |
| | i.p. | | | 0 | 50 |
| | | Learning | Rat | ↓ | 76 |

**Table 4.3 (continued)**

| *Peptide* | *Route* | *Effect* | *Species* | *Change* | *Ref.* |
|---|---|---|---|---|---|
| $\beta$-E | icv. | Seizure | Gerbil | ↑ | 52 |
| | | Temperature | Mouse | ↑ | 53 |
| | i.p. | Locomotor activity amphetamine-induced | Rat | ↓ | 54 |
| $\gamma$-E | s.c. | Startle reflex | | ↑ | 55 |
| | | Stereotypy, amphetamine-induced | | ↑ | 56 |
| Enk(Leu) | i.v. | Blood pressure, heart rate | Dog | ↓ | 57 |
| Enk(Met) | icv. | Temperature | Mouse | ↑ | 58 |
| | i.p. | Learning | Rat (foetus) | ↓ | 59 |
| Gastrin 1-17 | ic v. | Rumination | Sheep | ↓ | 60 |
| | i.v. | | | 0 | 60 |
| GRP | icv, i.p. | Satiety | Rat | ↓ | 17 |
| Gastrin | cns. | Drinking | | ↓ | 32 |
| Gastrin human 1-77 | i.p. | Satiety | | ↑ | 61 |
| Gastrin, penta and tetra | icv.,i.v. | Rumination | Sheep | ↓ | 60 |
| Glucagon | i.p. | Feeding | Rat | ↓ no satiety effect | 62 |
| GRF | cns. | Neuronal firing | Rat | ↑ predominantly | 63 |
| | icv.,i.v. | Gastrointestinal motility | Dog | ↑ duration | 64 |
| LHRH | cns. (preoptic area) | Neuronal activity | Rat | ↓ majority | 65 |
| | s.c. | Libido | Human (male) | ↓ | 66 |
| *Litorin* | *icv.* | Electroencephalogram | Rabbit | complex, activating | 16 |
| | | Gastric acid secretion | Rat | ↑ | 67 |
| | i.p. | Drinking | | ↑ | 19 |
| MIF-1 | i.m. | Aggression | Mouse | ↑ | 68 |
| | i.p. | Analgesia, morphine-induced | Rat | delayed onset | 69 |
| | | Tonic immobility | Lizard | ↑ | 70 |
| | s.c. | Dopamine receptor super-sensitivity, haloperidol-induced | Rat | ↑ | 71 |
| MSH | icv. | Grooming | | ↓ | 72 |

| | | | | | |
|---|---|---|---|---|---|
| MSH | cns. | Fever, LP-induced | Rabbit | ↑ | 73 |
| | icv. | Temperature | | ↑ | 74 |
| | i.p. | Aggression | Mouse (paired) | ↓ | 75 |
| MSH/ACTH-4-10 | s.c. | Attention | | ↓ latency | 77 |
| | s.c. | Avoidance, passive | | | |
| NPY | cns. | Blood pressure | Rat | ↓ | 78 |
| | cns. | Feeding | | ↑ | 79 |
| | (paraventricular nucleus) | | | | |
| | i.c.v. | | | ↑ | 80 |
| NT | i.p. | Blood pressure | Chicken | ↑ | 81 |
| | cns. | Dopamine release | Rat | ↑ | 82 |
| | (substantia nigra) | | | | |
| | | Dopamine metabolism (globus pallidus, striatum) | | ↓ 20 hours | 83 |
| | icis. | Emesis | Dog | ↓ | 84 |
| | icv. | Temperature | Rat | ↑ | 85 |
| | i.v. | Ventilation | | 0 | 24 |
| | | Drinking | | ↓ | 86 |
| OT | cns. | Heart rate | | ↑ | 87 |
| | (Dorsal motor nucleus of the vagus) | | | | |
| | icv. | Analgesia | | ↓ | 88 |
| | i.p. | Startle response | | ↓ | 89 |
| | s.c. | Tolerance to ethanol | | ↑ | 90 |
| Pancreatic PP | i.v . | Satiety | Dog | 0 | 91 |
| Sauvagine | icv. | Locomotor activity | Rat | ↓ | 43 |
| SP | | Depression, haloperidol-induced | | Reversed | 92 |
| | i.m. | Aggression | Cichlid Fish | Redirected | 93 |
| | i.v . | Blood pressure | Dog | ↑ | 94 |
| SP,SP7-11 | i.p.,3V | Locomotor activity | Mouse | ↑ | 95 |
| SRIF | cns. | Neuronal firing | Rat | ↑ predominantly ↓ some | 63 |
| | icv. | Ventilation | | 0 | 24 |
| SRIF-28 | icis. | Temperature | | ↑ | 96 |

**Table 4.3 (continued)**

| *Peptide* | *Route* | *Effect* | *Species* | *Change* | *Ref.* |
|---|---|---|---|---|---|
| SRIF-28 | icis. | Gastric secretion | Rat | ↓ | 15 |
| | icv. | Alcohol withdrawal tremor | | 0 | 98 |
| | | Electrical activity (hippocampus) | | ↓ (theta rhythm) | 98 |
| | | Locomotor activity | | ↓ | 99 |
| | | Piloerection | | ↓ | 99 |
| | | Shaking | | ↓ | 51 |
| VIP | it. | Blood pressure | | ↓ | 100 |
| | icv. | Sleep, pentobarbitol-induced | | ↑ | 100 |
| | | REM time | Cat | ↓ | 101 |
| | i.p. | Locomotor activity | Rat | ↑ | 61 |
| VP | hypothalamic graft | Drinking | Brattelboro rat | ↑ | 102 |
| VP antiserum | cns. | Memory, passive avoidance | Rat | ↑ | 103 |
| | icv. | Avoidance passive, retention | | ↑ latency | 77 |
| VP (Arg) | | Memory | Chicks (2 days old) | ↓ | 104 |
| | icis. | Gastric secretions | Rat | ↑ | 15 |
| | s.c., i.v. | Temperature | Gerbil | ↑ | 105 |
| VP (Lys) | i.n. | Conditioned avoidance response | Humans | ↓ acquisition ↓ extinction | 106 |
| | i.m. | Memory processing | | 0 | 107 |
| VP (hypothalamic graft to Brattelboro rat) | | Drinking, antidiuresis | Brattelboro rat | ↑ | 108 |
| VP | s.c. | Alcohol drinking | | ↑ | 109 |
| VT (Arg) | icv., i.p. | Learned response | Chicks | ↓ | 110 |
| | i.n. | Sleep, REM | Human (boys) | ↓ amount<br>↑ latency | 111 |
| | i.n. | Sleep, REM | Human narcoleptics | ↓ amount<br>↑ latency | 112 |
| | i.p. | Discrimination, aversive | Rat | ↓ reversal | 113 |
| | i.p. | Eye opening | Kitten | ↑ | 114 |
| VT | icv. | Blood pressure, heart rate | Rat | ↓, ↓ | 115 |

**Table 4.4** Effects of peptide administration on various behaviour patterns (modified according to Zadina *et al.*, 1986[1]; Woods *et al.*, 1986[122] and Figlewicz *et al.*, 1987[123]. Data are derived from studies in which peptides have been peripherally administered, unless differently indicated in brackets

| *Peptide* | *Effect* | *Reference* |
|---|---|---|
| **Alcohol intake** | | |
| BBS | decrease | 116 |
| Litorin | decrease | 116 |
| VP(Lys) | decrease | 117 |
| **Drinking** | | |
| Angiotensin II (icv.) | increase | 12 |
| BBS(icv) | increase or decrease, dependency on species | 18 |
| $\beta$-E (icv.) | 0 | 19 |
| CCK-8 (icv.) | decrease | 118 |
| DM (icv.) | increase or decrease | 18 |
| Litorin | decrease | 19 |
| MIF-1 | decrease | 120 |
| Motilin | decrease | 100 |
| NPY(icv.) | increase | 79 |
| NT | increase | 85 |
| PYY (cns.) | increase | 78 |
| Ranatensin (icv.) | decrease | 17 |
| Tachykinins (icv.) | dependency on species | 18 |
| TRH | increase | 121 |
| VP | decrease | 107 |
| **Food intake** | | |
| BBS | decrease | 122, 123 |
| Caerulein | decrease | 122, 123 |
| CCK | decrease | 122, 123 |
| CGRP | decrease | 124 |
| Galamin (cns) | increase | 125 |
| Glucagon | decrease | 122, 123 |
| GRP | decrease | 122, 123 |
| Insulin (cns.) | decrease | 122, 123 |
| NPY | increase | 122, 123 |
| Pancreatic PP (cns.) | increase | 122, 123 |
| PYY | increase | 122, 123 |
| SRIF | decrease | 122, 123 |
| TRH | decrease | 122, 123 |
| **Aggression** | | |
| MIF-1 | decrease | 68 |
| SP | decrease | 92 |
| **Sleep** | | |
| $\beta$-E | Slow-wave increase | 128 |
| Ceruletide | Hexobarbitol-induced increase | 35 |
| DM | Slow-wave increase | 128 |
| Muramyl dipeptide | 0 | 127 |
| VT(Arg) | REM increase amount, decrease latency | 111 |

In the last few years a particular attention has been focused on the ability of peptides to affect the higher integrative functions of the brain. It has been shown that systemic administration of VP(Arg)[129], VT(Arg)[110], Enk(methionine)[59], MSH, MSH/ACTH analogues[76], as well as VP analogue DDAVP[130] enhance the process of learning, both in the adult and developing animal. MSH, MSH/ACTH[76] and DDAVP[131] analogues may increase the *attention* when administered systemically, while β-endorphin decreases the attention[132]. The *avoidance* response is affected by many peptides including VP(Arg)[133], CCK–8[134], MSH, DDAVP[135] and MSH/ ACTH analogues[136], while *memory processing* may be influenced by the administration of VP(Arg), VP(Lys) and DDAVP[129], DSIP[137], MSH[76] and Enk(Met)[59] (Table 4.3).

## Peptides and vegetative functions of the central nervous system

Most peptides, when administered either systemically or applied locally, induce changes in the vegetative functions (Table 4.3) in parallel with other effects on the central nervous system. Although some of these effects could be attributed to the *peripheral action* of peptides, for the majority the effects on vegetative functions are mediated at least partly via their *central action*. For example, i.c.v. administration of VP causes an increase in *blood pressure* and *heart rate*[138] and the same effect can be obtained by iontophoretic c.n.s. application of VP(Arg)[138]. Intrathecal administration of TRH[139], as well as the central application of this peptide into the hypothalamic nucleus and preoptic nucleus, increases both the heart rate and blood pressure. Local application of TRH into nucleus tractus solitarius decreases the arterial blood pressure[140]. Intra-arterial and i.v. injection of both natural enkephalins increase the blood pressure and heart rate[141], and the same effects may be induced by the local c.n.s. application of the synthetic enkephalin analogues, DADLE and DAGO[142].

For some peptides the opposite effects on the same vegetative function were observed, depending on the route of administration. For example, oxytocin applied by iontophoresis into the dorsal vagal nucleus, produced a decrease in the heart rate[143], while i.c.v. injection of the same peptide had the opposite effect[138]. Intrathecal injection of oxytocin, on the other hand, decreased the heart rate, but increases the arterial blood pressure[139].

Systemic administration of angiotensin II[13], Enk (leucine)[144] and TRH[145] increased the blood pressure, while the same route for the administration of bradykinin[146], CRF[147], eledoisin, SP and VIP[146], induced a drop in the arterial blood pressure. Some peptides, like parathyroid hormone and NT, are ineffective with respect to the regulation of blood pressure[146].

*Gastric secretion* seems to be also under peptidergic control. For example, i.c.v. administration of BBS[148], calcitonin and CGRP[149], gastrin-related peptide and litorin[148], as well as intracisternal administration of BBS,

$\beta$-endorphin and gastrin-related peptides[149], all reduced gastric secretion. In contrast, the iontophoretic application of TRH into the dorsal vagal nucleus[143], or i.c.v. injection of TRH or its analogue[32], stimulated gastric secretion.

*Respiration* is stimulated by the ventricular administration of BBS, calcitonin and TRH[151], while the application of enkephalin analogue DADLE into the fourth ventricle produced a depressive effect on ventilation[152]. In contrast, the iontophoretic application of DADLE and DAGO into the nucleus tractus solitarius stimulated respiration[142].

*Body temperature* is affected by many peptides including angiotensin II and III[153], BBS, $\beta$-endorphin, gastrin related peptide, MSH, NT, SS, TRH[154], VP(Arg)[105], ACTH, MSH, litorin-like peptide[155], MIF-1[156], and CCK[157].

## Synthetic analogues of naturally occurring peptides

Many naturally occurring peptides are unstable in body fluids, since they are often subject to rapid enzymatic degradation in the blood and CSF, as well as by the brain tissue itself. The half-life for most peptides is different in body fluids from that in the central nervous system both with respect to the rate of degradation in these tissues and with regard to the chemical structure of the peptide molecule.

Over the past few years a considerable effort has been made by the pharmaceutical industry to produce analogues of peptide molecules which are resistant to enzymic degradation and yet retain their biological activity. This approach has produced synthetic molecules which have not only an extended half life in body fluids but often have a greatly enhanced activity, which is of especial value when these peptides are given systemically. For instance, the synthetic analogue ACTH/MSH (ORG 2766) given i.p. reduces both seizures and the uptake of morphine by the brain in experimental animals more effectively than ACTH itself[158]. Similarly, the synthetic analogues of BBS are more potent than BBS itself in reducing gastric secretion and lowering blood pressure[159].

Dynorphin A (fragment 1–13) is an extraordinarily potent opioid peptide[160], while [Nlc$^4$, D-Phe$^7$]-$\alpha$-MSH is reported to be 26 times more potent than $\alpha$-MSH in the adenylate cyclase assay[161]. [pGlu$^4$, Cyt$^6$, Arg$^8$]VP$^9$ is a more potent neuropeptide than VP(Arg) with selective effects on memory and related processes[162]. On the other hand, some of the peptide synthetic analogues may act as antagonists or specific inhibitors such as saralasin (a human type analogue of angiotensin II), [Sar$^1$, Val$^5$, Ala$^8$]-angiotensin II[163], [D-Phe$^7$]-bradykinin[164]. These examples are by no means exhaustive.

Synthetic analogues have been manufactured for many naturally occurring peptides including ACTH/MSH, bradykinin, casomorphin, CCK, caereulin, ceruletide, dermorphin, DSIP, opioid peptides, MSH, SS, TRH, VIP and VP, and some of them are given in Table 4.5. The table provides the

**Table 4.5** Synthetic analogues of naturally occurring peptides

| *Peptides and analogues* | *Amino acid sequence* | *Action* |
|---|---|---|
| **ACTH (human)** | Ser-Tyr-Ser-Met-Glu-His-Phe-Arg-Trp-Gly-Lys-Pro-Val-Gly-Lys-Lys-Arg-Arg-Pro-Val-Lys-Val-Tyr-Pro-Asn-Gly-Ala-Glu-Asp-Glu-Ser-Ala-Glu-Ala-Phe-Pro-Leu-Glu-Phe | |
| [Met(O)$^4$D-Lys$^8$-Phe$^9$] ACTH(4-9) ORG 2766 | Met[O]-Glu-His-Phe-D-Lys-Phe | Potent ACTH/MSH analogue |
| **Angiotensin II** | Asp-Arg-Val-Tyr-Ile-His-Pro-Phe | |
| [Val$^5$]-angiotensin II | Asp-Arg-Val-Tyr-Val-His-Pro-Phe | Hypertensive |
| [Sar$^1$]-angiotensin II | Sar-Arg-Val-Tyr-Ile-His-Pro-Phe | Hypertensive |
| [Sar$^1$,Ala$^8$]-angiotensin II | Sar-Arg-Val-Tyr-Ile-His-Pro-Ala | Inhibition to angiotensin II |
| [Sar$^1$,Ile$^8$]-angiotensin II | Sar-Arg-Val-Tyr-Ile-His-Pro-Ile | Specific inhibitor for angiotensin I |
| [Sar$^1$,Val$^5$,Ala$^8$]-angiotensin II (Saralasin) | Sar-Arg-Val-Tyr-Val-His-Pro-Ala | Specific inhibitor for angiotensin II |
| **Bradykinin** | Arg-Pro-Pro-Gly-Phe-Ser-Pro-Phe-Arg | |
| [D-Phe$^7$]-bradykinin | Arg-Pro-Pro-Gly-Phe-Ser-D-Phe-Phe-Arg | Competitive antagonist of bradykinin |
| Des-Arg$^9$-bradykinin | Arg-Pro-Pro-Gly-Phe-Ser-Pro-Phe | Biologically active bradykinin analogue |
| Des-Arg$^9$,[Leu$^8$]-bradykinin | Arg-Pro-Pro-Gly-Phe-Ser-Pro-Leu | Bradykinin inhibitor |
| **Dynorphin A** | Tyr-Gly-Gly-Phe-Leu-Arg-Arg-Ile-Arg-Pro-Lys-Leu-Lys-Trp-Asp-Asn-Gln | |
| Dynorphin A 1-13 | Tyr-Gly-Gly-Phe-Leu-Arg-Arg-Ile-Arg-Pro-Lys-Leu-Lys | Extraordinarilly potent opioid action |
| **$\beta$-E** ($\beta$-LPH 61-91) | Tyr-Gly-Gly-Phe-Met-Thr-Ser-Glu-Lys-Ser-Gln-Thr-Pro-Leu-Val-Thr-Leu-Phe-Lys-Asn-Ala-Ile-Ile-Lys-Asn-Ala-His-Lys-Lys-Gly-Gln | |
| **Enk(Leu)** | Tyr-Gly-Gly-Phe-Leu | |
| D-Ala$^2$-D-Leu$^5$-Enk(DADLE) | Tyr-D-Ala-Gly-Phe-Leu | Potent opioid action |
| Des-Tyr$^1$-Enk(Leu) | Gly-Gly-Phe-Leu | Tetrapeptide important for study of enkephalinase inhibition |

**Table 4.5 (continued)**

| *Peptides and analogues* | *Amino acid sequence* | *Action* |
|---|---|---|
| ACTH (human) | Ser-Tyr-Ser-Met-Glu-His-Phe-Arg-Trp- | |
| [D-Ser[5]]Enk(Leu)-Thr(DSLET) | Tyr-D-Ser-Gly-Phe-Leu-Thr | Specific $\delta$ opioid receptor agonist |
| **Enk(Met)** | Tyr-Gly-Gly-Phe-Met | |
| D-Ala[2]-Enk(Met)(DAME) | Tyr-D-Ala-Gly-Phe-Met | Potent Enk(Met) analogue |
| D-Ala[2]-Enk(Met)amide(DAMA) | Tyr-D-Ala-Gly-Phe-Met-NH | Potent Enk(Met) analogue |
| [D-Ala[2],N-MePhe[4],Gly[5]-ol]-Enk(DAGO) | Tyr-D-Ala-Gly-N-Me-Phe-Gly-ol | Selectively binds to $\mu$ opioid receptors |
| **$\alpha$-MSH** | *N*-acetyl-Ser-Tyr-Ser-Met-Glu-His-Phe-Arg-Trp-Gly-Lys-Pro-Val-$NH_2$ | |
| [Nle4-D-Phe7]-$\alpha$-MSH | *N*-acetyl-Ser-Tyr-Ser-Nle-Glu-His-D-Phe-Arg-Trp-Gly-Lys-Pro-Val-$NH_2$ | Potent in adenylate cyclase assay |
| **VP(Arg)** | Cys-Tyr-Phe-Gln-Asn-Cys-Pro-Arg-Gly-$NH_2$ | |
| [deamino-Cys[1]-D-Arg[8]VP (desmopressin) | 3-Mercaptopropionyl-Tyr-Phe-Gln-Asn-Cys-Pro-D-Arg-Gly-$NH_2$ | Specific antidiuretic and memory enhancement activity |
| [$\beta$-mercapto-$\beta$,$\beta$,cyclopenta-methylene-propionyl]′,O-Me-Tyr[2],Arg[8]-VP | [1-Mercaptocyclohexyl]-acetyl-Tyr [Me]-Phe-Gln-Asn-Cys-Pro-Arg-Gly-$NH_2$ | Antagonist of VP(Arg) |
| [pGlu[4],Cyt[8]]-VP-(Arg)4-9 | pGlu-Asn-Cys(Cys)-Pro-Arg-Gly-$NH_2$ | potent analogue of VP(Arg) with selective effects on memory |
| des-Gly-$NH_2$[9]-VP(Arg)[8] (DGAVP) | Cys-Tyr-Phe-Gln-Asn-Cys-Pro-Arg | Potent analogue of VP(Arg) |

amino acid sequences of peptides and their synthetic analogues with abbreviated nomenclature and related actions. Probably the best studied examples are the naturally occurring pentapeptides Enk (leucine) and Enk (methionine) and their synthetic analogues DADLE, DAGO, DAMA and DAME, which illustrate that the replacement of the amino acid L-alanine or D-alanine in a second position, or L-leucine by D-leucine in a fifth position of the molecule significantly increases the resistance of the molecule towards enzymic degradation yet at the same time preserving or potentiating its analgesic properties. Similar effects may be obtained by the replacement of methionine by methionine-amide at the fifth position in Enk (methionine). Such a change in the molecular structure of the peptide may significantly reduce the rate of its inactivation by altering the substrate affinity of the peptidase–peptide interaction.

## PEPTIDES AS POSSIBLE NEUROTRANSMITTERS AND/OR NEUROMODULATORS

The list of peptides as candidates for *neurotransmitters* or for *modulating* neurotransmission in the central nervous system is extending daily with novel names complemented by schematic illustrations of the neurons and circuits that contain and may receive the newly recognized peptidergic signals. So far as the criteria for neurotransmission are concerned, it has been generally accepted that peptides fulfil some of the rather rigorous proof for the synaptic transmitter status, ascribed for the classical amino acid and monoamine transmitters such as $\gamma$-aminobutyrate, dopamine, glutamate, aspartate, glycine or taurine[165]. In addition, we may also point out that many of the criteria for the synaptic transmitter status were established on the basis of studies with acetylcholine at the neuromuscular junction, which may not be relevant to other substances as candidates for neurotransmitters. A long history of central nervous system transmitter research has suggested that the substance must satisfy a large number of requirements in order to be considered as a *neurotransmitter*. The most rigorous of these requirements are: (a) to be contained in the nerve terminals at a specific synapse, (b) to be released upon the presynaptic depolarization into the synaptic cleft, and (c) to exert the electrophysiological effects upon the postsynaptic neuron which must be identical to the action of firing of the presynaptic neuron. Apart from that, to the criteria which have been inherited from the classical neurotransmitters such as acetylcholine, biogenic amines and amino acids, we must add the list: (d) the existence of the specific enzyme to inactivate each neurotransmitter, and/or (e) inactivation of neurotransmitter by the re-uptake into the nerve endings of the presynaptic neuron, and finally[169], (f) proof of the synthesis of neurotransmitter in the brain *in situ*. It appears that the 'classical' neurotransmitters account for approximately 40% of the known synapses present in the brain[166].

If we look at the growing list of peptides proposed as transmitters in the central nervous system, we find that, for nearly all the potential peptide transmitters, there is not, as yet, a complete experimental set of the above-mentioned criteria, and in every case some of these criteria are missing. Peptides as possible transmitter factors may, for didactical reasons, be divided into two large groups, depending whether they were discovered before or after their presumptive biological actions were specified[167,168]. Thus in the group *'factor first–function later'* are peptides that were discovered from their chemical properties, but without having any specific neural function. In other words, these peptides were given their chemical names because it was from their chemical structure that they were identified as biological products of the brain, while their possible function was investigated later. These include for example: substance P, neuropeptide Y, PTH-27, etc. In the other group *'assay first–factor later'*, the development of a bioassay for an unknown regulator factor was the starting point for the purification–isolation process, which essentially represents the classical Starling approach. To this group belong the following: endogenous opioid peptides, oxytocin, vasopressin, angiotensin, cholecystokinin, gastrin, CRF, GHRF, SRIF etc.

It has been accepted that every protein or peptide product being expressed in the brain depends for its synthesis on the transcription of a specific piece of genetic information into the messenger RNA which governs the synthesis of the gene product. By using the powerful methods of the recombinant DNA technology Bloom and his colleagues were able to demonstrate some 30 000 messenger RNAs that were specific for the mammalian brain and not detectable in other major organs[169]. Thus a genetic analysis of the brain and the functional significance of this analysis represents a potent and significant strategy for *molecular discovery*, which is ideal for investigation of peptide molecular complexity. In particular, this technique can be used to study distinct molecular systems involved in brain peptide synthesis, storage, release and recognition by the target cells[169]. The development of these new molecular genetic techniques now imposes new cellular, subcellular and biochemical criteria for the proof of neurotransmitter peptide pro-hormones. Thus, it is reasonable to conclude that it is difficult to accept a minimum single set of criteria for peptide CNS neurotransmitters. This is especially true if one takes into account that most of the present criteria were derived from research on the classical neurotransmitters, and we now have to reconcile this approach with that of the new experimental methods based on the modern genetic strategies for transmitter discovery.

Most of the brain peptides have in common at least one criterion which is valuable in classifying them as neurotransmitters, namely that the compound should be localized in the nerve endings. As we will see in Chapter 5, the peptidergic pathways were recognized in the brain by using specific immunocytochemical methods and many such peptides were localized in the nerve endings, for example endogenous opioid peptides, TRH, CRF, LHRH, somatostatin, vasopressin, oxytocin, DSIP, insulin, angiotensin II, bombesin-like peptides, cholecystokinin, VIP, etc. However, it has not been possible so far to demonstrate for all these peptides that they are synthesized *in situ* in the brain. On the other hand, the *neuropeptide concept* as formulated by de Wied[170] stated: 'Neuropeptides are endogenous substances present in nerve cells which are involved in nervous system function. Neuropeptides are synthesized in large precursor proteins, and several are formed in the same precursor such as ACTH and $\beta$-endorphin in pro-opiomelanocortin (POMC). A cascade of processes evolve in peptidergic neurons to express the genetic information into biologically active neuropeptides. These processes control the quantities of neuropeptides synthesized as well as the nature of their biological activity, through size, form and derivatization of the end-product. In this way sets of neuropeptides with different, opposite and more selective properties are formed from the same precursor'.

**Table 4.6** Electrophysiological effects of peptides on the mammalian cell membrane coupled to an opening or closing of potassium channel (taken from Kelly (1982)[177]; North and Egan (1982)[178]; North (1986)[176]).

| *Opening* | *Closing* |
|---|---|
| Enk (via $\mu$ and $\delta$ receptors) | SP |
| SRIF | VIP |
| | SRIF (concentraction dependent) |
| | BBS |
| | NT |

It has been possible to demonstrate depolarization-induced release as an important transmitter criterion, for many peptides, for example calcium-dependent release of enkephalins from the rat brain synaptosomes depolarized by potassium[171]. In the lateral septum, vasopressin-containing vesicles were identified in the synapses and calcium-dependent vasopressin release was demonstrated by depolarizing stimuli[172]. TRH release from the hypothalamic synaptosomal preparations has been known for some time, and these examples are by no means exhaustive[173]. In addition, the peptide attachment to the specific postsynaptic receptors and induction of biological effects identical with those achieved by direct nerve stimulation, the so-called 'synaptic mimicry', has been demonstrated for many neuropeptides. For example, depolarization of the neurones in the lateral septum can be elicited by superfusion of the slices with $10^{-9}$ mol/l solution of vasopressin[174], which is certainly a physiological concentration and thus can have a functional significance. Similarly, the application of bombesin into the cell body layer of the pyramidal cells of the hippocampus induces excitatory depolarization in the CA1 region of hippocampal slices[175]. Also the application of LHRH to frog sympathetic ganglion cells stimulates the nerve-evoked effects that produce late slow EPSPs, and this activation can be blocked by the LHRH antagonists K39. The ionic mechanisms of the electrophysiological effects of peptides on the membranes of mammalian nerve cells are fundamentally similar to those of classical neurotransmitters[176], in other words the same sets of ion channels can be modified both by peptides and classical transmitters, acting via shared second messenger systems. Table 4.6 illustrates some electrophysiological effects of peptides on mammalian nerve cell membrane which are coupled to an opening or closing of the potassium channel and have been obtained by intracellular recording methods. Thus we may summarize, that the decision whether a given peptide is a neurotransmitter or 'neuromodulator' is probably more a question of semantics than the lack of complete experimental evidence. However, whatever classification is used, it does not rule out the important role of peptides in the regulation of various neuronal circuits in the central nervous system.

## Neuropeptide–'classical' neurotransmitter interactions

The co-existence of several classical neurotransmitters and neuropeptides in neurons[179,180] together with the complex organization of the neuronal circuits in the central nervous system make the neuropeptide–'classical neurotransmitter' interactions difficult to study. However, some successes have been obtained by application of the quantitative receptor autoradiography with computer image analysis, combined immunostaining, retrograde tracing techniques, conventional electrophysiological and biochemical techniques. Throughout these studies it has become apparent that the communications between nerve cells involve both direct excitation and inhibition of neuronal firing by neuropeptides and classical neurotransmitters. Some examples for the co-existence of possible storage sites for peptides and monoamines in nerve endings in different nervous system areas is given in Table 4.7. As indicated in the table, neurons may secrete two classes of substances which involve two distinct types of secretory organelle[180]. Usually storage sites for classical neurotransmitters are represented by the small vesicles, typical synaptic vesicles, while the peptides appear to be stored in the large, dense cored vesicles, considered to be equivalent to secretory organelles in endocrine cells[180]. Several other chemical messenger interactions also modulate the response properties and functional capacity of nerve cells. This modulation may occur at two levels: (a) the *receptor* site (primary level), which involves changes in activation of intracellular second messengers that regulate the phosphorylation of specific proteins by protein kinases, and (b) the *gene* site (secondary level) involving regulation of the genomic activity and expression of gene products involved in nerve cell function. We illustrate these different types of neuropeptide–neurotransmitter interactions with the following examples.

**Table 4.7** Co-localization in nerve endings of possible storage sites for peptides and monoamine classical neurotransmitters (according to Kreiger, 1986[180]).

| *Brain region* | *Neuropeptide* | *Classical neurotransmitters* |
|---|---|---|
| Medulla oblongata | SP | serotonin |
| | TRH | serotonin |
| | SP + TRH | serotonin |
| Sympathetic ganglia | SRIF | noradrenaline |
| Superior cervical ganglion | Enk | noradrenaline |
| Carotid body | Enk | dopamine |
| Ventral tegmental area | CCK | dopamine |
| Autonomic ganglia | VIP | acetylcholine |

There are a number of pharmacological, immunocytochemical and neurochemical experiments which suggest that the tridecapeptide, *neurotensin*, may interact with nigrostriatal and mesolimbic dopaminergic pathways[181]. Also behavioural experiments have revealed an activation of dopaminergic cells in midbrain following an intracerebral administration of neurotensin[182], and this peptide has also been found to modify the metabolism of dopamine in the central nervous system[183,184]. The highest concentrations of neurotensin were found in substantia nigra, ventral tegmental area, nucleus accumbens, striatum and frontal cortex, all brain regions known to be rich in dopaminergic cell bodies[185]. The most striking recent evidence for neurotensin–dopamine interactions was provided by means of an *in vitro* quantitative receptor autoradiography[186]. A close similarity is seen in the autoradiographic distribution of neurotensin binding sites and topographical localization of dopaminergic cell bodies. Furthermore, this data suggested that neurotensin receptors are localized on dopaminergic cell bodies and/or dendrites in the substantia nigra and ventral tegmental area, which has been taken as a morphological basis for neurotensin–dopamine interactions in these regions. Injection of 6-OH-dopamine into the midbrain tegmentum strongly reduced [$^{125}$I]neurotensin binding in the lateral and central portions of the caudate-putamen, but not in the nucleus accumbens, which supported the hypothesis that in the caudate-putamen, the majority of [$^{125}$I]neurotensin binding sites are localized presynaptically[186,187]. Investigations on humans have shown that chronic neuroleptic treatment enhances neurotensin binding in the substantia nigra[188]. In contrast, there is a dramatic loss of neurotensin binding in Parkinsons disease, which is characterized by the destruction of dopaminergic cell bodies in the substantia nigra[189]. These findings support the hypothesis that neurotensin receptors are present on dopaminergic cells, which suggests that this peptide may regulate the activity of dopaminergic neurons in man. On the other hand, it has been demonstrated in the rat that dopamine reduces the affinity and increases the number of [$^{3}$H]neurotensin binding sites in the subcortical limbic forebrain[190], and vice versa; neurotensin was shown to regulate dopaminergic receptors in different brain structures[191,192]. These findings, in the rat, support the hypothesis that the interactions between these substances may occur via mutual regulation of their respective receptors.

Vasoactive intestinal polypeptide (VIP) is another peptide whose interactions have been studied with classical neurotransmitters such as noradrenaline and serotonin. It has been reported that biochemical actions of VIP in the cortex involve its ability to stimulate cyclic AMP formation, even more potently than noradrenaline[193]. In addition, the distribution of VIP- and noradrenaline-containing circuits, as determined cytochemically, exhibits a contrasting, but complementary anatomy since both fibre systems may have the same target, the pyramidal cells[194]. Subsequent findings have shown that VIP and noradrenaline can act synergistically to increase cyclic AMP level in cerebral cortex[195]. Iontophoretic application of VIP during subthreshold stimulation induced by noradrenaline administration produced a pronounced inhibition of cellular discharge. The inhibition was independent of the effect of VIP prior to noradrenaline, showing, from the electrophysiological point of view, a significant interaction between these molecules[196]. Experiments on single neurons confirmed the synergistic inhibition of VIP and noradrenaline, although VIP alone may also elicit an excitatory effect on the test neuron[196]. It has been suggested from the electrophysiological interactions of VIP–noradrenaline that these molecules may act via a common biochemical event, namely the generation of cyclic AMP. This secondary messenger may mediate both noradrenaline – and VIP-evoked depression of neuronal firing in cortex, thus acting as the intracellular mediator of their synergistic interaction. As we pointed out previously, an important aspect of the receptor modulation, which can be easily observed by means of quantitative autoradiography, is the peptide-induced regulation of brain receptor sensitivity to classical neurotransmitters. It has been demonstrated that suprachiasmatic nucleus of the hypothalamus contains one of the densest serotonin innervation in the brain and this is where a significant number of VIP-containing nerve cell bodies have also been found[197]. Biochemical and morphological data indicate the possible functional interactions between VIP and serotonin in the suprachiasmatic nucleus[197,198], while direct incubation of VIP with [$^3$H]serotonin on sections of the suprachiasmatic nucleus has demonstrated the ability of the peptide to decrease the number of serotonin binding sites without affecting the affinity of binding[186], the effect being strictly localized to this brain region.

Significant interactions have also been demonstrated between somatostatin and acetylcholine. In electrophysiological experiments it has been shown that somatostatin has an inhibitory effect on the spontaneous firing rate in neurons unless acetylcholine was included in the multibarrel iontophoresis assembly[199]. Subsequent studies have shown that somatostatin, when applied alone, directly depresses the spontaneous discharge rate of neurons with a slow onset and offset of its effects. In contrast, somatostatin enhances the excitatory actions of acetylcholine but not those of glutamate, which suggests that the interaction is selective to acetylcholine[200]. However, the cellular and molecular mechanisms which underlie the interactions between somatostatin and acetylcholine require further study.

## PEPTIDES AND NEUROENDOCRINE AXIS

Many peptides and neurotransmitters are involved in the control of the secretion of various pituitary hormones. The morphological and functional basis of this regulatory activity depends mainly on the *multifactorial* organization of communications between the median eminence and hypothalamic signalling system on the one hand and the pituitary on the other. It has been shown that several neuropeptides may interact presynaptically with each other in the median eminence and other regions of the hypothalamus, modulating each other's efficacy at the level of membrane transduction mechanisms. These findings have contributed to the so-called concept of the *coding system of hypothalamic information* for the control of pituitary hormones secretion[201]. According to this concept, the neuroendocrine controlling structures are organized in a complex network, in which neurosecretory neurons containing peptidergic and classical transmitters are closely interconnected by numerous collateral axons within this network. Some signals may carry the *biological message* itself, while others are involved in *message addressing*, to subselect targets sharing common expression of receptors for the same elementary signals. We will illustrate this concept of a new design in neuroendocrine systems with a few examples, in which neuropeptides play an important role.

It has been shown in many studies that significant interactions exist between the major coupling chains of pituitary receptors and their different transduction modes (Figure 4.1). Since it has been accepted that for any given receptor, the receptor coupling is the same in all tissues, it appears that the major difference in cell responses depends upon cell-specific and selective rearrangements of the coupling chains themselves, and not upon different modes of reception in the elementary signals. Thus, for instance, VIP and CRF may interact with the stimulatory component of the *GTP binding coupling protein complex*, while dopamine and SRIF interact with the inhibitory component of the same complex, which controls the adenylate cyclase activity and formation of the cAMP. On the other hand, in the same pituicyte, LHRH and TRH activate the phospholipase A which stimulates the formation of arachidonic acid and its metabolites, the leukotrienes and prostaglandins. Also, activation of the phospholipase C via stimulation of the phospholipase $A_2$ increases the level of prostaglandins which may act synergically with GTP binding coupling protein complex on adenylate cyclase, modulating its stimulation[202,203]. This may amplify an initial phospholipid-dependent message mediated via diacylglycerol (DAG). The feedback mechanism is bidirectional since cyclic nucleotides (cAMP and cGMP) may inhibit

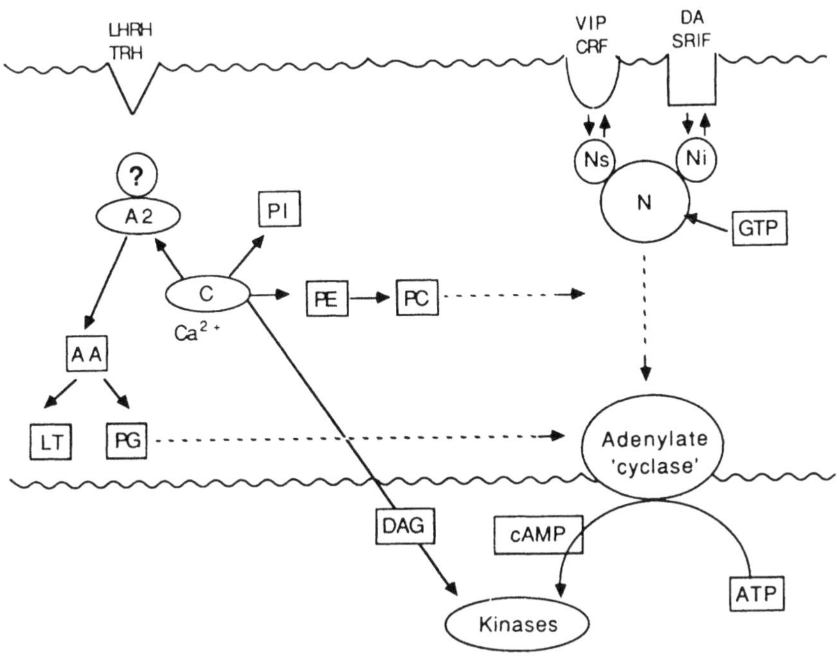


**Figure 4.1** Schematic representation of the interactions between the major coupling chains of the pituitary receptors (according to Kordon *et al.* (1987)[201]. Ni, Ns, inhibitory and stimulatory components of N (GTP binding coupling protein complex); A2, C, phospholipase A2 and C; AA, LT, PG, arachidonic acid and its metabolites leukotrienes and prostaglandins); DAG, diacylglycerol; PI, phosphoinositol; PE, PC, phosphatidylethanolamine and -choline

phospholipase C, shutting off calcium mobilization and terminating the message initiated by cyclase coupling[204]. Feedback mechanisms also exist between cAMP- or DAG-dependent protein kinases on the one hand, and phosphorylation of receptors or coupling proteins on the other[205]. This interaction may be achieved either in a *homologous* manner, when receptor phosphorylation is self-regulated, or in a *heterologous* manner when receptors can modulate sensitivity or coupling of another receptor.

Interchain interactions are involved in the pulsatile control of LH secretion, and it has been shown that cAMP can potentiate the pituitary response to subsequent LHRH stimulation, the effect being mediated via phospholipase C[206]. Stimulation of adenylate cyclase by subproducts of phospholipase $A_2$ activation, via cAMP, can feedback on phospholipase C, and this may explain why artificial stimulation of LH secretion by cAMP alone is almost ineffective. Several neuronal elements are involved in morphofunctional organization of the hypothalamic loops responsible for the episodic control of LH and GH secretion. It has been demonstrated that pulsatile control systems for LH and GH can operate in the absence of any extrinsic input to the mediobasal hypothalamus[207], but that dopamine, $\beta$-endorphin and

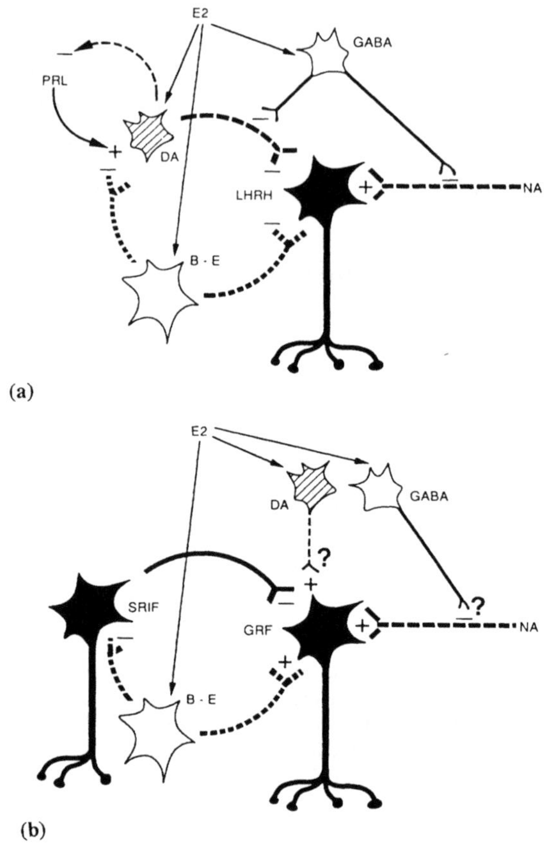


**Figure 4.2** Schematic representation of the hypothalamic neuronal circuits involved in the pulsatile control of (a) LH and (b) GH secretion (according to Kordon *et al.*(1987)[201]. Dopamine (DA), β-endorphin (β-E) and LHRH neurons exhibit redundant interactions for LH control; β-E, SRIF and GRF neurons show a comparable organization pattern in the case of GH control. Noradrenergic (NA) inputs modulate the activity of both systems, with an added GABA regulation (only demonstrated so far for LHRH); prolactin is able to interfere with the DA component of LH regulation. Several neuronal elements of both loops are targets of estradiol (E2)

LHRH neurons exhibit redundant interactions for LH control, while β-endorphin, SRIF and GRF neurons show a comparable organization pattern for the GH control (Figure 4.2). For example, dopamine or its agonists may block LH pulsatility in the rat[208], whereas opiate peptides can inhibit the presynaptic release of dopamine[209], LHRH and somatostatin[210]. On the other hand, they have an opposite effect on GRF release which is suggested by the experimental finding that opiate peptides cannot stimulate GH secretion after passive immunization with anti-GRF antisera[211,212]. It has been found also that GRF release and the resulting pulsatile GH secretory episodes are preceded by an intrahypothalamic drop in somatostatin release[213].

Apart from the above-mentioned neuronal circuits, in the multifactorial regulation of the pulsatile LH and GH secretion, an extrinsic noradrenergic component is also involved, and noradrenergic neurons can modulate the activity of both systems[214], while an additional GABA regulation has been demonstrated but at present only for LHRH[215]. Recently it has been discovered that several components of both loops including tuberoinfundibular dopamine and $\beta$-endorphin neurons and preoptic GABA interneurons can express estrogen receptors and are targets for estradiol ($E_2$)[216], which may account for the modulation of episodic frequencies by the steroid.

Another example of complex interactions between neuropeptides and hormones involved in signal transmission, is the role of *opiates* in stress-induced hormonal changes. Most of these changes are reflected in altered plasma levels of various pituitary hormones (prolactin, GH) and can be prevented by treatment with opiate antagonists, which strongly suggests that endogenous opioid peptides are involved in mediating stressful stimuli to neuroendocrine control systems[217]. The release of many hypothalamic neuropeptides may be blocked by opioid peptides[210] and it has been accepted that presynaptic inhibition of opiates may account, at least in part, for these effects. Since opiates also exert similar effects on neuropeptide release in other brain structures, it appears that an increased opiate tone may induce a temporary upheaval of the normal regulatory circuits in the brain[201].

Although in the *stress-response* the hypothalamo–hypophyseo–adrenal axis is activated in different ways, depending on the stress, the adrenal cortex has been recognized for a long time as the major organizer of the non-specific adaptive syndrome[218]. However, more recent investigations suggest that a variety of neuropeptides is involved in modulating the neuroendocrine state during physiological response to the stress[219,220]. These investigations were conducted in an attempt to improve understanding of the different pathways involved in stress-induced ACTH release, depending on whether the nature of stress is systemic or neural[221]. Among several hypothalamic substances, CRF has been identified as the most predominant component in the release of ACTH from the anterior pituitary cells. However, in addition to CRF, the other neuropeptides, such as vasopressin, oxytocin, angiotensin II, VIP, PTH-27 as well as biogenic amines, may induce ACTH secretion from adenohypophysis and are also involved in the mechanisms of the response to stress[222–225]. It has been found that as well as the direct action of stress on CRF neurons, stress may also act on vasopressinergic, oxytocynergic and/or angiotensin II containing neurons, causing the release of these neuropeptides, which can then either potentiate the action of CRF at the pituitary level, or act synergistically with CRF on the target cells[224–226].

Several neuropeptides and classical neurotransmitters are involved in control of prolactin secretion from the adenohypophysis. It has been shown that major peptides and transmitters may interact by modulating each other's presynaptic release or by affecting each other's coupling at the level of the lactotroph membrane, or affecting directly the multifactorial regulation of the pituitary cells. Among neuropeptides an important role has been attributed to opioid peptides that can modulate dopamine-dependent cyclase inhibition which correlates with PRL secretion from lactotrophs, most likely at the coupling level[227]. On the other hand these interactions may provide a double control system to the gland, for example VIP can simultaneously temper the release of PRL, by inhibiting SRIF at the median eminence, as well as stimulating secretion of PRL by acting directly on the pituitary[201]. TRH and NT may also influence the secretion of prolactin at the level of the pituitary.

## Receptor-mediated effects of peptides in cerebral microvessels

It has been demonstrated that isolated cerebral microvessels contain peptide receptors some of which are linked to the *adenylate cyclase–cyclic AMP system*, such as VIP[228], PTH[229] and angiotensin II[230], which stimulate adenylate cyclase activity increasing cyclic AMP levels in the endothelial cells, which then can trigger a cascade of metabolic signals without the peptide actually crossing the blood–brain barrier. Thus, for example, a physiological concentration of bovine PTH (1-34) increases capillary adenylate cyclase activity in a manner similar to the actions of $\beta$-adrenergic agents (isoproterenol) or forskolin (Table 4.8). This may explain the possible neurotoxic effects of PTH, particularly in conditions such as renal failure when plasma levels of this peptide are greatly increased[231].

**Table 4.8** Adenylate cyclase activity of the isolated cerebral bovine microvessels in the presence of homologous PTH and $\beta$-adrenergic agents (according to Pardridge (1986)[282]).

| | *Additive* | *Adenylate cyclase activity (% of basal)* |
|---|---|---|
| PTH1-34 | 5 nmol/l | 115 |
| | 10 nmol/l | 142 |
| | 100 nmol/l | 175 |
| Isoproterenol | 100 μmol/l | 170 |
| Forskolin | 100 μmol/l | 256 |

Basal activity was linear for up to 20 min and averaged 14.2±0.7 pmol $min^{-1}$ $mg^{-1}$. Peptide and $\beta$-adrenergic agents were incubated with micro vessels into [$^{32}P$]ATP containing medium for 10 min at 30 °C.

Other peptides stimulate *polyphosphoinositide* metabolism in the endothelial cells, leading to the formation of *inositol triphosphate* which then mobilizes *calcium* from the intracellular stores, as in the case of bradykinin[232] and vasopressin[233], which exert their effects on the cerebral microvessels via bradykinin-B and vasopressin-V receptors. It has been suggested that the

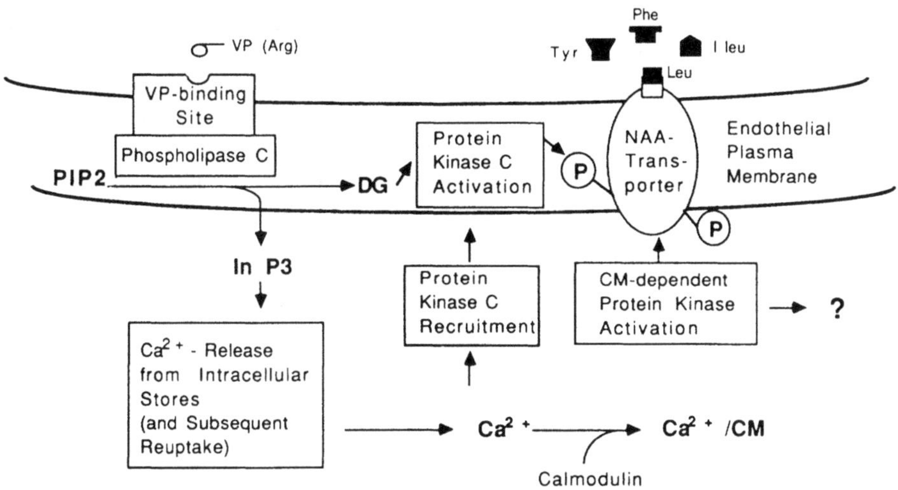


**Figure 4.3** Signal transduction events following VP(Arg) interactions with the cerebral endothelial $V_1$-receptors (according to Ermisch *et al.*(1988)[237]. Abbreviations: CM, calmodulin; DG, diacylglycerol; InP3, inositol 1,4,5-triphosphate; NAA, neutral amino acid: $PIP_2$, phosphotidylinositol 4,5-bisphosphate

peptide-induced activation of the cellular messenger signals (cAMP, cGMP, phosphoinositides) take place irrespective of the peptide origin, which may derive either from brain or from blood, and without peptide crossing the blood–brain barrier from blood to brain, or from brain to blood[234].

Radioligand-binding studies in the isolated microvessels were performed with either $^{125}$I or $^{3}$H-labelled VP(Arg)[235], which demonstrate saturable kinetics, and a Scatchard analysis indicated only high affinity binding sites, with an apparent dissociation constant, $K_D$ of 2.7 nmol/l, and a maximal binding capacity of 390 fmol/mg protein. The number of binding sites was calculated to be within the range from 4000 to 20 000 per one hippocampal endothelial cell[236]. However, the method used did not distinguish between the anatomical polarization of the binding sites, with respect to the luminal and abluminal side[237]. Immunocytochemical studies revealed *vasopressinergic fibres* in close vicinity to the brain capillaries[238], which may suggest the abluminal binding of VP(Arg) to the brain microvessels. On the other hand, VP(Arg) injected in a physiological concentration into the internal carotid artery, together with a rapidly penetrating substrate, may influence the passage of substrate across the blood–brain barrier, which has been suggested to occur via receptors located on the luminal side of the capillary wall[236,239]. Thus, it can be concluded that VP(Arg) receptors may be localized both on the luminal and abluminal side of the blood–brain barrier[237].

Saturable binding of VP(Arg) has also been demonstrated *in vivo* by the i.v. injection 'integral' technique in the hippocampus[239]. However, a non-transport function of VP endothelial receptors has also been suggested, which involves cellular response and further *signal transduction* linked to the regulation of substrate transport (Figure 4.3)[237] The signal transduction in the hippocampal endothelium after an intracardial perfusion of a physiological solution containing $10^{-8}$ mol/l VP(Arg) consisted of an increase in $Ca^{2+}$ intracellular storage, which was made visible as electron dense deposits[240]. X-ray microanalysis confirmed that in fact these deposits contain more $Ca^{2+}$ than the deposit-free cytoplasm, which suggest the VP-induction of calcium re-uptake. The findings support the hypothesis that VP binding sites at the blood–brain barrier may represent $V_1$-receptors. However, we may note that involvement of VP(Arg) blood–brain barrier receptors in the signal transduction does not rule out the possibility that these receptors may be also involved in the transfer of VP across the endothelial cell[241].

Specific binding to the isolated microvessels has been also shown for some other small peptides such as *oxytocin*[242] and *angiotensin II*[230] as well as for a variety of large peptides and proteins including *insulin*, *insulin-like growth factors*, *transferrin* and *albumin*[243] whose functions will be explained in more detail in Chapter 7.

## PEPTIDE RECEPTORS AT THE CIRCUMVENTRICULAR ORGANS

Peptide interactions at the blood–CSF barrier are by no means restricted to the choroid plexus, but also include the circumventricular organs (CVO), i.e. a number of small areas situated in close vicinity of the ventricular surface, which like the choroid plexuses are vascularized by permeable fenestrated capillaries (see Chapter 1). These areas lack a blood–brain barrier, and it has been suggested that CVO's may be a transfer site for the movement of neuropeptides from one part of the brain to another. However it should be noted that these regions do have tight functions at the ependyma so direct access into the CSF may be restricted[244]. Several studies have provided evidence for the presence of peptide receptors in the circumventricular organs[245]. These include insulin, prolactin, angiotensin II, calcitonin and ACTH, suggesting that after binding to the nerve endings at the CVO and/or distribution in the interstitial space of the CVO, the circulating peptides may reach other brain regions by diffusion, or some other way of transfer.

Subsequent studies have characterized in more detail the possible function of CVO in neuropeptide brain homeostasis. For example, it has been demonstrated that $^{125}$I-insulin injected into the plasma is concentrated in neuronal endings in the CVO, and in some cases transported deep within the brain[245] and, vice versa, labelled insulin injected into the CSF is taken up by surrounding hypothalamic tissue around the third ventricle[246]. Thus, insulin has access to the brain at the CVO, and is also transported across the blood–brain barrier, where it enters the neural tissue directly or is taken up into the neural tissue from CSF[247]. A possible role of insulin receptors at the CVO would be to mediate the action of the rapid fluctuations of the plasma insulin level on the central nervous system. Specific receptors were also found for angiotensin II in the subfornical organ and organum vasculosum laminae terminalis, and it has been suggested that these CVO receptors may be sensitive to blood-borne angiotensin II[248]. Receptor-mediated transport has also been demonstrated for prolactin in the median eminence[249] which may possibly mediate sudden changes in prolactin plasma levels in the brain. Receptors for vasopressin, presumably $V_1$ type have been also identified in the regions of the central nervous system without a blood–brain barrier[239].

## OTHER FUNCTIONS OF PEPTIDES IN THE BRAIN

### Cerebral vasomotor effects

It has been demonstrated that many brain derived peptides such as somatostatin, bombesin, VIP, substance P, neurotensin and neuropeptide Y are closely associated with cerebral vessels, which has provided evidence for the morphological basis of possible peptide interactions with brain blood vessels[250–253]. Neuropeptides were found both in the peptidergic nerve terminals within the walls of brain blood vessels, and in the nerve endings which terminate on brain capillaries[234,254,255]. It has also been reported that peptides to some degree can pass the endothelial lining of cerebral vessels[241,243,256,257], which suggests the possibility that both brain-derived and blood-borne peptide may influence the cerebral circulation[234,243,253]. On the other hand, investigations of peptide interactions with brain capillaries indicate that it is most likely that they serve other functions than simply modulating cerebral blood-flow. For example peptides may be responsible for *short-term alterations* in brain capillary endothelial function, such as selective changes in the blood–brain barrier permeability properties, and *long-term* effects by modulation of the unique biochemical properties of the brain endothelial cell via their action on the gene expression[253]. In this chapter we have already discussed peptide interactions with cerebral capillaries and their induced signal transduction effects, while in the following chapter we will describe how large peptides and proteins interact with isolated cerebral microvessels. Here we will provide some evidence on the effect of peptides on the larger cerebral vessels, which may be of considerable importance when considering their possible modulatory role in the regulation of cerebral blood flow.

Most of the vasomotor effects of neuropeptides were studied on the isolated blood-vessel preparations taken from different species with an aim to elucidate their vasomotor action by studying the effects on the vascular smooth muscle of the brain *in vitro*[256]. Thus a variety of neuropeptides has been found which induce *vasoconstriction* in feline pial arteries. The concentration causing half maximal contraction ($ED_{50}$ value) for angiotensin II, VP(Arg), bradykinin and VP(Lys) had values ranging between 3.7 to 7.5

nmol/l, showing that these different molecules were almost equally potent in their dose-dependent constrictor action. Oxytocin and somatostatin were somewhat less potent with the $ED_{50}$ value between 16 and 30 $\mu$mol/l. On the other hand, no vasoconstriction could be observed with calcitonin, enkephalins, neurotensin, pentagastrin and substance P. The presence of calcium uptake inhibitors, nifedipine and nimodipine, in the organ bath considerably reduced the contraction induced by VP(Lys), bradykinin and oxytocin, which indicates that the entry of extracellular calcium is of importance for the vascular effect. The competitive antagonist for angiotensin II, [$Sar^1$, $Ala^8$]-angiotensin II, counteracted the constrictor effect of angiotensin II on cerebral vessels. When the same peptides were tested on extracranial arteries, no significant difference in potency was observed apart from the inability of VP(Arg) to induce any constrictor effect on these extra-brain vessels. Both enkephalins were found to *dilate* pial arteries in a dose-dependent manner with $ED_{50}$ being 39 and 31 $\mu$mol/l for Enk(Leu) and Enk(Met), respectively. Enkephalin-induced vasodilation could be blocked by the opiate receptor blocking agent, naloxone, while the blockade of the adrenergic, cholinergic and histaminergic receptors were without effect. Vasodilation did not occur with neurotensin, pentagastrin and substance P, and it may be concluded that these peptides are without effect on cerebral vessels *in vitro*.

VIP had no effect on the isolated cerebral blood vessels during resting conditions, but when an active tone was induced by either prostaglandin $F_{2\alpha}$ or serotonin, the cumulative application of VIP resulted in dose-dependent relaxation of both pial and extracranial vessels[258]. Intracarotid infusion of VIP ($10^{-11}$ moles/min) did not significantly alter cerebral blood flow, oxygen consumption[258] or EEG recording in the baboon[259]. On the other hand, when the same concentration of VIP was perfused following an intracarotid bolus injection of a hypertonic urea solution to open the blood–brain barrier, this peptide induced a significant increase, in both cerebral blood flow and oxygen consumption; these changes were accompanied by a moderate elevation of arterial blood pressure, and a consistent alteration in the EEG. The infusion of larger concentrations of VIP into the internal carotid artery of baboon ($10^{-10}$ moles/min) produced a small, but consistent increase in cerebral blood flow associated with an increased oxygen consumption, and both the blood flow and metabolism returned to the preinfusion level 5 min after the cessation of the VIP infusion[258].

It has been demonstrated *in vivo*, by means of the radioactive microsphere technique, that a specific $\delta$-opioid agonist (DSLET) produced an increase of the cerebral blood-flow in the hippocampus, brain stem, striatum and cerebellum in the rat, 2.5 min after an intracerebroventricular administration of this agonist in the hippocampus[260]. The effect was transient, and the time course was similar to that for triggering and maintaining the seizure phenomena induced by enkephalins.

## Effects on the CSF secretion

So far little information is available about peptide effects on CSF secretion, although continuous production of CSF represents an important source for physiological water homeostasis in the CNS. It has been shown that antidiuretic hormone (ADH) exerts an inhibitory effect on CSF secretion when presented to the choroid plexus from the CSF side during an artificial ventriculo-cisternal perfusion in the rabbit[261]. This inhibition of CSF formation has been accompanied by a significant reduction of Na entry from blood into the CSF. Similar effects may be obtained following an intracarotid injection of peptide[262]. An inhibitory effect of ADH has also been demonstrated on outflow of the cerebrospinal fluid during a vasogenic brain oedema induced by cold lesion in rabbit[263]. In addition, an increase of ADH was found during development of brain oedema in both CSF and peripheral blood, which, together with a decrease of CSF outflow from the cisterna magna, has been taken as proof that ADH may be one of the factors responsible for the extracellular fluid regulation during vasogenic type of brain oedema[263].

Some peptides have been shown to exert vasomotor effects on the choroid plexus blood vessels which may alter perfusion of these organs and so influence the rate of CSF secretion. For example, peptidergic VIP nerve terminals were found in the mammalian choroid plexus, and the cumulative application of VIP resulted in a dose-dependent relaxation of the choroidal blood vessels which had an active tone[264]. DSIP has also been found to induce a moderate dose-dependent inhibition of CSF secretion rate in an isolated perfused choroid plexus of the sheep[265].

## Peptides and alterations in the blood–brain barrier permeability

It has been suggested that the so-called *short-term regulation* neuropeptides may be responsible for temporary changes in the blood–brain barrier permeability by (a) altering the affinity and capacity of different carrier transport systems at the barrier, (b) by modulating the pinocytosis across the barrier, and (c) by opening the tight junctions between endothelial cells[243,253]. It has been demonstrated that VP(Arg) and oxytocin may enhance blood–brain barrier permeability to [$^3$H]orotic acid, the RNA precursor, when injected intracarotidally[266,267], but there was no effect on glucose extraction. In addition, VP(Arg) has been found to change blood–brain barrier transport properties to L-leucine, by decreasing both the $K_m$ and $V_{max}$ for the transport process[237,239,268,269]. These findings suggest a dose-dependent specific effect induced by the blood-borne peptide. The alterations in the amino acid transport are probably triggered by specific VP(Arg) binding to the cerebral capillary endothelium which may induce a specific transduction signal via $V_1$ receptors on the luminal surface of the endothelial cells, as shown in Figure 4.3[237]. Following the intracerebroventricular injection of VP(Arg) it has been reported that this peptide may alter the blood–brain barrier permeability to water and this will increase water uptake[270]; this effect may be mediated by noradrenergic pathways[271].

By the use of indicator diffusion method in man, it has been shown that large intravenous doses of insulin enhance the unidirectional influx of D-glucose across the blood–brain barrier[272,273]. Systemic administration of ORG 2766, a synthetic analogue of ACTH/MSH 4-9, has been shown to reduce the permeability of the blood–brain barrier to antipyrine[274].

The formation of endothelial cell junctions has been shown to be under the influence of the calcium–calmodulin system, as well as the second messengers of cAMP and cGMP[275–277]. Thus it has been postulated that transient changes in the permeability of the blood–brain barrier to plasma proteins may occur normally under physiological conditions. Neuropeptides secreted by the brain may also alter brain capillary protein phosphorylation and dephosphorylation, and this protein kinase activity in brain capillaries was found to be nearly as active as that from synaptosomal preparations[278]. The findings of an active blood–brain barrier phosphorylation–dephosphorylation enzyme has provided a strong biochemical basis for studying the rapid control of brain capillary function by brain peptides[253].

There is evidence that the carrier-mediated system capable of transporting small, *N*-tyrosinated peptides such as *N*-Tyr-MIF-1 and enkephalins, across the blood–brain barrier may be modulated by the intraventricularly administered leucine[279]. Kinetic analysis indicated that leucine produces a non-competitive type of modulation, probably by interacting with the carrier–ligand complex. The possibility is also suggested that disturbances in amino acid metabolism may produce some of their effects on the central nervous system by modulating the blood–brain barrier transport systems for peptides and possibly for other substances.

Apart from the short-term regulation of blood–brain barrier permeability, it has been suggested that peptides may also play a role in the *long-term* regulation of genetic expression in the brain capillary endothelium. In 1981 Stewart and Wiley[280] reported the results of an important paradigm related to the genetic regulation of the biochemical organization of the blood–brain barrier phenomena in the brain. Namely, these authors transplanted a quail brain before vascularization into the embryonic chick coelum, and found that quail brain transplant was vascularized by vessels of brain type, with the morphological and biochemical characteristics of the blood–brain barrier. Tests were performed and these capillaries were shown to exclude the trypan blue and had few pinocytotic vesicles. When the reverse experiment was performed, namely the quail somites, before vascularization of the tissue, were transplanted to embryonic chick brain, the muscle transplant was subsequently vascularized by the vessels of muscle type, which lacked the typical biochemical and morphological characteristics of the blood–brain barrier. These vessels were now actively pinocytotic, and did not exclude trypan blue. These experiments unequivocally emphasized that specific biochemical and structural features of the blood–brain barrier are induced during maturation by some trophic factors secreted by the brain, the most serious candidates for that process being neuropeptides of gial origin[243,253]. Identification of the blood–brain barrier specific proteins such as $\gamma$-glutamyl transpeptidase, 50–55kDa triplet of phosphoproteins and/or low-molecular weight proteins in the range of 14kDa to 18kDa[281], and 46kDa protein[243], that are not found in brain cells or in synaptosomal plasma membrane fractions, have opened a new potential area for studying the hypothesis of peptide induction of the blood–brain barrier[243].

## REFERENCES


1. Zadina, J.E., Banks, W.A. and Kastin, J.E. (1986). Central nervous system effects of peptides 1980–1985. *Peptides*, **7**, 497–537
2. Schacter, M. (1981). Enkephalins and endorphins. *Br. J. Hosp. Med.*, **25**, 128–136
3. Vilberg, T.R., Panksepp, J., Kastin, A.J. and Coy, D.H. (1984). The pharmacology of endorphin modulation of chick distress vocalization. *Peptides*, **5**, 823–827
4. Lipkowski, A.W., Konecka, A.M. and Sroczynska, I. (1982). Double-enkephalins-synthesis, activity on guinea-pig ileum, and analgesic effect. *Peptides*, **3**, 697–700
5. Honde, C. and Bueno, L. (1984). Evidence for central neuropeptidergic control of rumination in sheep. *Peptides*, **5**, 81–83
6. Kastin, A.J., Olson, R.D., Schally, A.V. and Coy, D.H. (1979). CNS effects of peripherally administered peptides. *Life Sci.*, **25**, 401–414
7. Aloyo, V.J., Spruijt, B., Zwiers, H. and Gispen, W.H. (1983). Peptide-induced excessive grooming in the rat: The role of opiate receptors. *Peptides*, **4**, 833–836
8. Saint-Come, C., Acker, G.R. and Strand, F.L. (1982). Peptide influences on the development and regeneration of motor performance. *Peptides*, **3**, 439–449
9. Blair, R., Galina, Z.H., Sutherland, C. and Amit, Z. (1983). ACTH 1-39 but not naltrexone produces biphasic effects on locomotor activity. *Peptides*, **4**, 117–120
10. Thornhill, J.A. and Saunders, W.S. (1984). Thermoregulatory (core, surface and metabolic) responses of unrestrained rats to repeated POAH injections of beta-endorphin or adrenocorticotropin. *Peptides*, **5**, 713–719
11. Appel, N.M. and Van Loon, G.R. (1983). Activation of angiotensin II receptors in brain potentiates the stimulating effect of endogenous opioid neurons on central sympathetic outflow. *Peptides*, **4**, 59–62
12. Miceli, M.O. and Malsbury, C.W. (1983). Feeding and drinking responses in the golden hamster following treatment with cholecystokinin and angiotensin II. *Peptides*, **4**, 103–106
13. Gross, P.M., Kadekaro, M., Andrews, D.W., Sokoloff, L. and Saavedra, J.M. (1985). Selective metabolic stimulation of the subfornical organ and pituitary neural lobe by peripheral angiotensin II. *Peptides*, **6** (Suppl. 1), 145–152
14. Widerlov, E., Mueller, R.A., Frye, G.D. and Breese, G.R. (1984). Bombesin increases dopamine function in rat brain areas. *Peptides*, **5**, 523–528
15. Tache, Y., Vale, W., Rivier, J. and Brown, M. (1981). Brain regulation of gastric acid secretion in rats by neurogastrointestinal peptides. *Peptides*, 2 (Suppl. 2), 51–55
16. Tartara, A., Bo, P. and Savoldi, F. (1982). Neuropeptides and cerebral electric activity in rabbits. *Peptides*, 3, 125–127
17. DeCaro, G., Massi, M., Micossi, L.G. and Perfumi, M. (1984). Drinking and feeding inhibition by ICV pulse injection or infusion of bombesin, ranatensin and litorin to rats. *Peptides*, **5**, 607–613

18. DeCaro, G., Massi, M., Micossi, L.G. and Perfumi, M. (1982). Angiotensin II antagonists versus drinking induced by bombesin or eledoisin in pigeons. *Peptides*, **3**, 631–636
19. Gibbs, J., Kulkosky, P.J. and Smith, G.P. (1981). Effects of peripheral and central bombesin on feeding behavior of rats. *Peptides*, **2** (Suppl. 2), 179–183
20. Kulkosky, P.J., Gray, L., Gibbs, J. and Smith, G.P. (1984). Feeding and selection of saccharin after injections of bombesin, LiCl and NaCl. *Peptides*, **2**, 61–64
21. Tenner, T.E.J., Yang, C.M., Chang, J.K., Schimizu, M. and Pang, P.K.T. (1980). Pharmacological comparison of bPTH-(1-34) and other hypotensive peptides in the dog. *Peptides*, **1**, 285–288
22. Twery, M.J. and Moss, R.L. (1985). Calcitonin and calcitonin gene-related peptide alter the excitability of neurons in rat forebrain. *Peptides*, **6**, 373–378
23. Hughes, J.J., Levine, A.S., Morley, J.E., Gosnell, B.A. and Silvis, S.E. (1984). Intraventricular calcitonin gene-related peptide inhibits gastric acid secretion. *Peptides*, **5**, 665–667
24. Niewoehner, D.E., Levine, A.S. and Morley, J.E. (1983). Central effects of neuropeptides on ventilation in the rat. *Peptides*, **4**, 277–281
25. Giusti, P., Carrara, M., Zampiron, S., Cima, L. and Borin, G. (1985). Are calcitonins analgesic and/or hyperalgesic? *Peptides*, **6** (Suppl. 3), 277–282
26. Candeletti, S., Romualdi, P., Spadaro, C., Spampinato, S. and Ferri, S. (1985). Studies on the antinociceptive effect of intrathecal salmon calcitonin. *Peptides*, **6** (Suppl. 3), 273–276
27. Motta, M. (1985). Neuroendocrine effects of some amphibian peptides. *Peptides*, **6** (Suppl. 3), 131–135
28. Hsiao, S. and Wang, C.H. (1983). Continuous infusion of cholecystokinin and meal pattern in the rat. *Peptides*, **4**, 15–17
29. Ruiz-Gayo, M., Duage, V., Menant, I., Begue, D., Gacel, G. and Rogues, B.P. (1985). Synthesis and biological activity of BOC[Nle 28, NLE 31]CCK 27–33, a highly potent CCK 8 analogue. *Peptides*, **6**, 415–420
30. Dumbrille-Ross, A. and Seeman, P. (1984). Dopamine receptor elevation by cholecystokinin. *Peptides*, **5**, 1207–1212
31. Katsuura, G., Itoh, S. and Hsiao, S. (1985). Specificity of nucleus accumbens to activities related to cholecystokinins in rats. *Peptides*, **6**, 91–96
32. Willis, G.L., Hansky, J. and Smith, G.C. (1984). The role of some central catecholamine systems in cholecystokinin-induced satiety. *Peptides*, **5**, 41–46
33. Itoh, S. and Katsuura, G. (1985). Fronto-cortical regulation of beta-endorphin actions in the rat. *Peptides*, **6**, 237–240
34. Cohen, S.L., Knight, M., Tamminga, C.A. and Chase, T.N. (1983). Tolerance to the antiavoidance properties of cholecystokinin-octapeptide. *Peptides*, **4**, 67–70
35. Zetler, G. (1982). Ceruletide, ceruletide analogues and cholecystokinin octapeptide (CCK-8): effects on motor behavior, hexobarbital-induced sleep and harman-induced convulsions. *Peptides*, **3**, 701–704
36. Stacher, G., Steinringer, H., Schmierer, G., Schneider, C. and Winklehner, S. (1982). Cholecystokinin octapeptide decreases intake of solid food in man. *Peptides*, **3**, 133–136
37. Smith, G.P., Kulkosky, J.P. and Simansky, K.J. (1984). Ceruletide acts in the abdomen, not in the brain, to produce satiety. *Peptides*, **5**, 1149–1157
38. Stacher, G., Steinringer, H., Schmierer, G., Winklehner, S. and Schneider, C. (1982). Ceruletide increases threshold and tolerance to experimentally induced pain in healthy man. *Peptides*, **3**, 955–962
39. DeCastiglione, R. (1981). Structural requirements of ceruletide-like peptides for activity on gut and brain. *Peptides*, **2** (Suppl. 2), 61–63
40. Fargeas, M.J., Fioramonti, J. and Bueno, L. (1985). Calcitonin gene-related peptide: brain and spinal action on intestinal motility. *Peptides*, **6**, 1167–1171
41. Eberly, L.B., Dudley, A.A. and Moss, R.L. (1983). Iontophoretic mapping of corticotropin-releasing factor (CRF) sensitive neurons in the rat forebrain. *Peptides*, **4**, 837–841

42. Kalin, N.H., Shelton, S.E., Kraemer, G.W. and McKinney, W.T. (1983). Corticotropin-releasing factor administered intraventricularly to Rhesus monkeys. *Peptides*, **4**, 217–220
43. Negri, L., Noviello, L. and Noviello, V. (1985). Effects of sauvagine, urotensin I and CRF on food intake in rats. *Peptides*, **6** (Suppl. 3), 53–57
44. Eaves, M., Thatcher-Britton, K., Rivier, J., Vale, W. and Koob, G.F. (1985). Effects of corticotropin releasing factor on locomotor activity in hypophysectomized rats. *Peptides*, **6**, 923–926
45. Graf, M., Zadina, J.E. and Schoenenberger, G.A. (1982). Amphetamine-induced locomotor behavior of mice is influenced by DSIP. *Peptides*, **3**, 729–731
46. Olson, G.A., Roig-Smith, R., Mauk, M.D., LaHoste, G.J., Coy, D.H., Hill, C.W. and Olson, R.D. (1981). Differential effects of neuropeptides on short-term memory in primates. *Peptides*, **2** (Suppl. 1), 131–136
47. Ren, M.F., Lu, C.H. and Han, J.S. (1985). Dynorphin-A-(1-13) antagonizes morphine analgesia in the brain and potentiates morphine analgesia in the spinal cord. *Peptides*, **6**, 1015–1020
48. Petrie, E.C., Tiffany, S.T., Baker, T.B. and Dahl, J.L. (1982). Dynorphin (1-13): analgesia, hypothermia, cross-tolerance with morphine and beta-endorphin. *Peptides*, **3**, 41–47
49. De Caro, G., Massi, M., Micossi, L.G. and Perfumi, M. (1982). Angiotensin II antagonists versus drinking induced by bombesin or eledoisin in pigeons. *Peptides*, **3**, 631–636
50. Vilberg, T.R., Panksepp, J., Kastin, A.J. and Coy, D.H. (1984). The pharmacology of endorphin modulation of chick distress vocalization. *Peptides*, **5**, 823–827
51. Tan, D. and Tsou, K. (1985). Differential motor and blood pressure effects of intrathecal oxytocin and TRH in the rat. *Peptides*, **6**, 1191–1193
52. Bajorek, J.G. and Lomax, P. (1982). Modulation of spontaneous seizures in the mongolian gerbil: effects of beta-endorphin. *Peptides*, **3**, 83–86
53. Bloom, A.S. and Tseng, L. (1981). Effects of beta-endorphin on body temperature in mice at different ambient temperatures. *Peptides*, **2**, 293–297
54. Yehuda, S., Zadina, J., Kastin, A.J. and Coy, D.H. (1980). D-amphetamine-induced hypothermia and hypermotility in rats: changes after systemic administration of beta-endorphin. *Peptides*, **1**, 179–185
55. Hoehler, F.K. and Sandman, C.A. (1981). Effects of alpha-MSH and beta-endorphin on startle reflex in rat. *Peptides*, **2** (Suppl. 1), 137–141
56. Kiraly, I., Tapfer, M., Borsy, J. and Graf, L. (1981). Further evidence for the neuroleptic-like activity of $\gamma$-endorphin. *Peptides*, **2**, 9–12
57. Giles, T.D. and Sander, G.E. (1983). Mechanism of the cardiovascular response to systemic intravenous administration of leucine-enkephalin in the conscious dog. *Peptides*, **4**, 171–175
58. Sakurada, T., Sakurada, S., Watanabe, S., Matsumura, H., Kisara, K., Akutsu, Y., Sasaki, Y. and Suzuki, K. (1983). Actions of intracerebroventricular administration of kyotorphin and an analog on thermoregulation in the mouse. *Peptides*, **4**, 859–863
59. Stickrod, G., Kimble, D.P. and Smotherman, W.P. (1982). Met-enkephalin effects on associations formed *in utero*. *Peptides*, **3**, 881–883
60. Honde, C. and Bueno, L. (1984). Evidence for central neuropeptidergic control of rumination in sheep. *Peptides*, **5**, 81–83
61. Olson, R.D., Kastin, A.J., von Almen, T.K., Coy, D.H. and Olson, G.A. (1980). Systemic injections of gastro-intestinal peptides alter behavior in rats. *Peptides*, **1**, 383–385
62. Geary, N. and Smith, G.P. (1982). Pancreatic glucagon fails to inhibit sham feeding in the rat. *Peptides*, **3**, 163–166
63. Twery, M.J. and Moss, R.L. (1985). Sensitivity of rat forebrain neurons to growth hormone-releasing hormone. *Peptides*, **6**, 609–613
64. Bueno, L., Fioramonti, J. and Primi, M.P. (1985). Central effects of growth hormone-releasing factor (GRF) on intestinal motility in dogs: involvement of dopaminergic receptors. *Peptides*, **6**, 403–407

65. Rothfield, J.M., Carstens, E. and Gross, D.S. (1985). Neuronal responsiveness to gonadotropin-releasing hormone and its correlation with sexual receptivity in the rat. *Peptides*, **6**, 603–608
66. Ehrensing, R.H., Kastin, A.J. and Schally, A.V. (1981). Behavioral and hormonal effects of prolonged high doses of LHRH in male impotency. *Peptides*, **2** (Suppl. 1), 115–121
67. Guglietta, A., Strunk, C.L., Irons, B.J. and Lazarus, L.H. (1985). Central neuromodulation of gastric acid secretion by bombesin-like peptides. *Peptides*, **6** (Suppl. 3), 75–81
68. Teskey, G.C. and Kavaliers, M. (1985). Prolyl-leucyl-glycinamide reduces aggression and blocks defeat-induced opioid analgesia in mice. *Peptides*, **6**, 165–167
69. Datta, P.C., Sandman, C.A. and Hoehler, F.K. (1982). Attenuation of morphine analgesia by alpha-MSH, MIF-1, melatonin and naloxone in the rat. *Peptides*, **3**, 443–437
70. Cashner, F.M., Olson, R.D., Erickson, D.G. and Olson, G.A. (1981). Effects of MIF-1 and sex differences on tonic immobility duration in the lizard *Anolis carolinensis*. *Peptides*, **2** (Suppl. 1), 161–165
71. Chiu, P., Rajakumar, G., Chiu, S., Johnson, R.L. and Mishra, R.K. (1985). Mesolimbic and striatal dopamine receptor supersensitivity: prophylactic and reversal effects of L-prolyl-L-leucyl-glycinamide (PLG). *Peptides*, **6**, 179–183
72. O'Donohue, T.L., Handelmann, G.E., Chaconas, T., Miller, R.L. and Jacobowitz, D.M. (1981). Evidence that N-acetylation regulates the behavioral activity of alpha-MSH in the rat and human central nervous system. *Peptides*, **2**, 333–334
73. Glyn-Ballinger, J.R., Bernardini, G.L. and Lipton, J.M. (1983). Alpha-MSH injected into the septal region reduces fever in rabbits. *Peptides*, **4**, 199–203
74. Lipton, J.M. and Glyn, J.R. (1980). Central administration of peptides alters thermoregulation in the rabbit. *Peptides*, **1**, 15–18
75. Nowell, N.W., Thody, A.J. and Woodley, R. (1980). The source of an aggression-promoting olfactory cue, released by alpha-melanocyte stimulating hormone, in the male mouse. *Peptides*, **1**, 69–72
76. Sandman, C.A., Beckwith, B.E. and Kastin, A.J. (1980). Are learning and attention related to the sequence of amino acids in ACTH/MSH peptides? *Peptides*, **1**, 277–280
77. Wimersma Greidanus, T.J.B. van (1980). MSH/ACTH 4-10: a tool to differentiate between the role of vasopressin in memory consolidation or retrieval processes. *Peptides*, **3**, 7–11
78. Carter, D.A., Vallejo, M. and Lightman, S.L. (1985). Cardiovascular effects of neuropeptide Y in the nucleus tractus solitarius of rats: relationship with noradrenaline and vasopressin. *Peptides*, **6**, 421–425
79. Stanley, B.G., Daniel, D.R., Chin, A.S. and Leibowitz, S.F. (1985). Paraventricular nucleus injections of peptide YY and neuropeptide Y preferentially enhance carbohydrate ingestion. *Peptides*, **6**, 1205–1211
80. Levine, A.S. and Morley, J.E. (1984). Neuropeptide Y: a potent inducer of consummatory behavior in rats. *Peptides*, **5**, 1025–1029
81. Carraway, R. and Bhatnagar, Y.M. (1980). Isolation, structure and biological activity of chicken intestinal neurotensin. *Peptides*, **1**, 167–174
82. Myers, R.D. and Lee, T.F. (1983). *In vivo* release of dopamine during perfusion of neurotensin in substantia nigra of the unrestrained rat. *Peptides*, **4**, 955–961
83. Napier, T.C., Gay, D.A., Hulebak, K.L. and Breese, G.R. (1985). Behavioral and biochemical assessment of time-related changes in globus pallidus and striatal dopamine induced by intranigrally administered neurotensin. *Peptides*, **6**, 1057–1068
84. Wu, M., Harding, K., Hugenholtz, H. and Kucharczyk, J. (1985). Emetic effects of centrally administered angiotensin II, arginine vasopressin and neurotensin in the dog. *Peptides*, **6** (Suppl. 1), 173–175
85. Martin, G.E., Bacino, C.B. and Papp, N.L. (1980). Hypothermia elicited by the intracerebral microinjection of neurotensin. *Peptides*, **1**, 333–339
86. Stanley, B.G., Hoebel, B.G. and Leibowitz, S.F. (1983). Neurotensin: effects of hypothalamic and intravenous injections on eating and drinking in rats. *Peptides*, **4**, 493–500

87. Rogers, R.C. and Hermann, G.E. (1985). Dorsal medullary oxytocin, vasopressin, oxytocin antagonist, and TRH effects on gastric acid secretion and heart rate. *Peptides*, **6**, 1143–1148
88. Kordower, J.H. and Bodnar, R.J. (1984). Vasopressin analgesia: specificity of action and non-opioid effects. *Peptides*, **5**, 747–756
89. King, M.G., Brown, R. and Kusnecov, A. (1985). An increase in startle response in rats administered oxytocin. *Peptides*, **6**, 567–568
90. Pucilowski, D., Kostowski, W. and Trzaskowska, E. (1985). The effect of oxytocin and fragment (MIF-1) on the development of tolerance to hypothermic and hypnotic action of ethanol in the rat. *Peptides*, **6**, 7–10
91. Levine, A.S., Sievert, C.E., Morley, J.E., Gosnell, B.A. and Silvis, S.E. (1984). Peptidergic regulation of feeding in the dog *(Canis familiaris)*. *Peptides*, **5**, 675–679
92. Jolicoeur, F.B., Rondeau, D.B., Belanger, F., Fouriezos, G. and Barbeau, A. (1980). Influence of substance P on the behavioral changes induced by haloperidol in rats. *Peptides*, **1**, 103–107
93. Christ, H. (1985). Effects of MET-ENK, substance P and SRIF on the behavior of *Hemichromis bimaculatus*. *Peptides*, **6**, 139–148
94. Sander, G.E., Giles, T.D. and Rice, J.C. (1985). Cardiovascular interactions between methionine-enkephalin and substance P in the conscious dog. *Peptides*, **6**, 133–137
95. Hall, M.E. and Stewart, J.M. (1984). Modulation of isolation-induced fighting by N- and C-terminal analogs of substance P: evidence for multiple recognition sites. *Peptides*, **5**, 85–89
96. Wakabayashi, I., Tonegawa, Y. and Shibasaki, T. (1983). Hyperthermic action of somatostatin-28. *Peptides*, **4**, 325–330
97. Huffman, L.J., Campbell, G.T. and Gilmore, J.P. (1983). Renal function and pituitary hormone release during cerebral osmostimulation and TRH in dogs. *Peptides*, **4**, 843–847
98. Drust, E.G. and Crawford, I.L. (1983). Comparison of the effects of TRH and D-Ala2-Metenkephalinamide on hippocampal electrical activity and behavior in the unanesthetized rat. *Peptides*, **4**, 239–243
99. Katsuura, G., Yoshikawa, K., Itoh, S. and Hsiao, S. (1984). Behavioral effects of thyrotropin releasing hormone in frontal decorticated rats. *Peptides*, **5**, 899–903
100. Itoh, S., Katsuura, G. and Yoshikawa, K. (1985). Hypermotility induced by vasoactive intestinal peptide in the rat; its reciprocal action to cholecystokinin octapeptide. *Peptides*, **6**, 53–57
101. Drucker-Colin, R., Bernal-Pedraza, J., Fernandez-Cancino, F. and Oksenberg, A. (1984). Is vasoactive intestinal polypeptide (VIP) a sleep factor? *Peptides*, **5**, 837–840
102. Gash, D. and Sladek, J.R. (1980). Vasopressin neurons grafted into Brattleboro rats: viability and activity. *Peptides*, **1**, 11–14
103. Greidanus, T.B.V.W. and Veldhuis, H.D. (1985). Vasopressin: site of behavioral action and role in human mental performance. *Peptides*, **6** (Suppl. 2), 177–180
104. Davis, J.L., Pico, R.M. and Cherkin, A. (1983). Prolyl-L-arginyl-glycineamide induces memory enhancement in chicks. *Peptides*, **4**, 401–404
105. Lee, R.J. and Lomax, P. (1983). Thermoregulatory, behavioral and seizure modulatory effects of AVP in the gerbil. *Peptides*, **4**, 801–805
106. Ehrensing, R.H., Michell, G.F. and Baker, R.P. (1982). Vasopressin's effects on acquisition and extinction of conditioned avoidance response to smoking. *Peptides*, **3**, 527–530
107. Fehm-Wolfsdorf, G., Born, J., Elbert, T., Voigt, K. and Fehm, H.L. (1985). Vasopressin does not enhance memory processes: a study in human twins. *Peptides*, **6**, 297–300
108. Gash, D., Sladek, C.D. and Sladek, J.R.J. (1980). A model system for analyzing functional development of transplanted peptidergic neurons. *Peptides*, **1** (Suppl. 1), 125–134
109. Myers, R.D., Critcher, E.C. and Cornwell, N.N. (1983). Effect of chronic vasopressin treatment on alcohol drinking of Brattleboro HZ and DI rats. *Peptides*, **4**, 359–366
110. Davis, J.L. and Pico, R.M. (1984). Arginine vasotocin delays extinction of a conditioned avoidance behavior in neonatal chicks. *Peptides*, **5**, 1221–1223

111. Pavel, S., Goldstein, M., Petrescu, M. and Popa, M. (1981). REM sleep induction in prepubertal boys by vasotocin: evidence for the involvement of serotonin containing neurons. *Peptides*, **2**, 245–250
112. Pavel, S., Goldstein, R. and Petrescu, M. (1980). Vasotocin, melatonin and narcolepsy: possible involvement of the pineal gland in its patho-physiological mechanism. *Peptides*, **1**, 281–284
113. Beckwith, B.E, and Tinius, T.P. (1985). Vasopressin and vasotocin facilitate reversal of a brightness discrimination. *Peptides*, **6**, 383–386
114. Goldstein, R. (1984). The involvement of arginine vasotocin in the maturation of the kitten brain. *Peptides*, **5**, 25–28
115. Zerbe, R.L., Kirtland, S., Faden, A.I. and Feuerstein, G. (1983). Central cardiovascular effects of mammalian neurohypophysial peptides in conscious rats. *Peptides*, **4**, 627–630
116. Kulkosky, P.J., Roque, M. and Sanchez, M.R. (1985). Bombesin and litorin inhibit ethanol intake. *Peptides*, **6** (Suppl. 2), 103–105
117. Rigter, H. and Crabbe, J.C. (1985). Vasopressin and ethanol preference. I. Effects of vasopressin and the fragment DGAVP on altered ethanol preference in Brattleboro diabetes insipidus rats. *Peptides*, **6**, 669–676
118. Denbow, D.M. and Myers, R.D. (1982). Eating, drinking and temperature responses to intracerebroventricular cholecystokinin in the chick. *Peptides*, **3**, 739–743
119. Deviche, P. and Schepers, G. (1984). Intracerebroventricular injection of ostrich beta-endorphin to satiated pigeons induces hyperphagia but not hyperdipsia. *Peptides*, **5**, 691–694
120. Olson, R.D., Kastin, A.J., Olson, G.A., King, B.M., von Almen, T.K., Berzas, M.C., Ibanez, M.L. and Coy, D.H. (1980). MIF-1 suppresses deprivation-induced fluid consumption in rats. *Peptides*, **1**, 353–357
121. Frye, G.D., Luttinger, D., Nemeroff, C.B., Vogel, R.A., Prange, A.J. and Breese, G.R. (1981). Modification of the actions of ethanol by centrally active peptides. *Peptides*, **2** (Suppl. 1), 99–106
122. Woods, S.C., Taborsky, C.J. and Porte, D.T. (1986). Central nervous system control of nutrient homeostasis. *Handbook Physiol.*, Section 1, Volume IV, 365–411 (Am. Physiol. Soc.)
123. Figlewicz, D.P., Lacour, F., Sipols, A., Porte, D.T. and Woods, S.C. (1987). Gastroenteropancreatic peptides and the central nervous system. *Ann. Rev. Physiol.*, **49**, 383–395
124. Tannenbaum, C.S. and Coltzman, D. (1985). Calcitonin gene-related peptide mimics calcitonin actions in brain on growth hormone release and feeding. *J. Endocrinol.*, **116**, 2685–2687
125. Kyzkouli, S.E., Stanley, B.G. and Leibowitz, S.F. (1985). Galanin: stimulation of feeding induced by medial hypothalamic injection of this novel peptide. *Psychopharmacol. Bull*, **21**, 412–418
126. Brown, R. and King, M.G. (1984). Arginine vasotocin and aggression in rats. *Peptides*, **5**, 1135–1138
127. Fornal, C., Markus, R. and Radulovacki, M. (1984). Muramyl dipeptide does not induce slow-wave sleep or fever in rats. *Peptides*, **5**, 91–95
128. Nistico, G., Bagetta, G. and De Sarro, G.B. (1985). Behavioral and spectrum power effects of opioid peptides in chicks. *Peptides*, **6** (Suppl. 3), 137–141
129. Hoffman, P.L., Szabo, G. and Tabakoff, B. (1988). The effects of vasopressin and related peptides on tolerance to ethanol. In: *Peptide and Amino Acid Transport Mechanisms in Central Nervous System*, eds., Lj. Rakic, D.L. Begley, H. Davson and B.V. Zlokovic (Macmillan, London) 147–156
130. Beckwith, B.E., Petros, T., Kanaan-Beckwith, S., Couk, D.I. and Haug, R.J. (1982). Vasopressin analog (DDAVP) facilitates concept learning in human males. *Peptides*, **3**, 627–630
131. Beckwith, B.E., Couk, D.I. and Till, T.S. (1983). Vasopressin analog influences the performance of males on a reaction time task. *Peptides*, **4**, 707–709

132. Hoehler, F.K. and Sandman, C.A. (1981). Effects of alpha-MSH and beta-endorphin on startle reflex in rat. *Peptides*, **2** (Suppl. 2), 137–141
133. Hoffman, P.L. (1987). Central nervous system effects of neurohypophysical peptides. In: *The Peptides Vol. 8*, ed., C.W. Smith (Academic Press, New York), 239–295
134. Cohen, S.L., Knight, M., Tamminga, C.A. and Chase, T.N. (1983). Tolerance to the antiavoidance properties of cholecystokinin-octapeptide. *Peptides*, **4**, 67–70
135. Hamburger-Bar, R., Klein, A. and Belmaker, R.H. (1985). The effect of chronic vs. acute injection of vasopressin on animal learning and memory. *Peptides*, **6**, 23–25
136. Wimersma Greidanius, T.J.B. van. (1982). MSH/ACTH 4–10: A tool to differentiate between the role of vasopressin in memory consolidation or retrieval processes. *Peptides*, **3**, 7–11
137. Susic, V., Masirevic, G. and Totic, S. (1987). The effects of delta sleep inducing peptides (DSIP) on wakefulness and sleep patterns in the cat. *Brain Res.*, **414**, 262–270
138. Zerbe, R.L., Kirtland, S., Faden, A.I. and Feuerstein, G. (1983). Central cardiovascular effects of mammalian neurohypophysical peptides in conscious rats. *Peptides*, **4**, 627–630
139. Tan, D. and Tsou, K. (1985). Differential motor and blood pressure effects of intrathecal oxytocin and TRH in the rat. *Peptides*, **6**, 1191–1193
140. Feuerstein, G., Hassen, A.H. and Faden, A.I. (1983). TRH: cardiovascular and sympathetic modulation in brain nuclei of the rat. *Peptides*, **4**, 617–620
141. Varagic, V.M., Stojanovic, V. and Dzoljic, E. (1988). The effect of enkephalins and enkephalinase inhibitors on the central cholinergic mechanisms participating in the peripheral adrenergic activation. In: *Peptide and Amino Acid Transport Mechanisms in the Central Nervous System*. eds., Lj. Rakic, D. Begley, H. Davson and B.V. Zlokovic, (Macmillan, London), 157–166
142. Hassen, A.H., Feuerstein, G. and Faden, A.I. (1983). Differential cardiovascular effects mediated by mu and kappa opiate receptors in hindbrain nuclei. *Peptides*, **4**, 621–625
143. Rogers, R.C. and Hermann, G.E. (1985). Dorsal medullary oxytocin, vasopressin, oxytocin antagonist, and TRH effects on gastric acid secretion and heart rate. *Peptides*, **6**, 1143–1148
144. Giles, T.D. and Sander, G.E. (1983). Mechanism of the cardiovascular response to systemic intravenous administration of leucine-enkephalin in the conscious dog. *Peptides*, **4**, 171–175
145. Huffman, L.J., Campbell, G.T. and Gilmore, J.P. (1983). Renal function and pituitary hormone release during cerebral osmostimulation and TRH in dogs. *Peptides*, **4**, 843–847
146. Tenner, T.E.J., Yang, C.M., Chang, J.K., Schimizu, M. and Pang, P.K.T. (1980). Pharmacological comparison of bPTH-(1-34) and other hypotensive peptides in the dog. *Peptides*, **1**, 285–288
147. Kalin, N.H., Shelton, S.E., Kraemer, G.W. and McKinney, W.T. (1983). Associated endocrine, physiological and behavioral changes in Rhesus monkeys after intravenous corticotropin-releasing factor administration. *Peptides*, **4**, 211–215
148. Guglietta, A., Strunk, C.L., Irons, B.J. and Lazarus, L.H. (1985). Central neuromodulation of gastric acid secretion by bombesin-like peptides. *Peptides*, **6** (Suppl. 3), 75–81
149. Tache, Y., Vale, W., Rivier, J. and Brown, M. (1981). Brain regulation of gastric acid secretion in rats by neurogastrointestinal peptides. *Peptides*, **2** (Suppl. 2), 51–55
150. Goto, Y. and Tache, Y. (1985). Gastric erosions induced by intracisternal thyrotropin-releasing hormone (TRH) in rats. *Peptides*, **6**, 153–156
151. Niewoehner, D.E., Levine, A.S. and Morley, J.E. (1983). Central effects of neuropeptides on ventilation in the rat. *Peptides*, **4**, 277–281
152. Holaday, J.W. (1982). Cardiorespiratory effects of $\mu$ and $\delta$ opiate agonists following third or fourth ventricular injections. *Peptides*, **3**, 1023–1029
153. Wilson, K.M. and Fregley, M.J. (1985). Factors affecting angiotensin II-induced hypothermia in rats. *Peptides*, **6**, 695–701

154. Morley, J.E., Levine, A., Oken, M.M., Grace, M. and Kneip, J. (1982). Neuropeptides and thermoregulation: the interactions of bombesin, neurotensin, TRH, somatostatin, naloxone and prostaglandins. *Peptides*, **3**, 1–6
155. Lipton, J.M. and Glyn, J.R. (1980). Central administration of peptides alters thermoregulation in the rabbit. *Peptides*, **1**, 15–18
156. Yehuda, S. and Kastin, A.J. (1980). Interaction of MIF-1 or alpha-MSH and d-amphetamine or chlorpromazine on thermoregulation and motor activity of rats maintained at different ambient temperatures. *Peptides*, **1**, 243–248
157. Denbow, D.M. and Myers, R.D. (1982). Eating, drinking and temperature responses to intracerebroventricular cholecystokinin in the chick. *Peptides*, **3**, 739–743
158. Goldman, H., Krasnewich, D., Murphy, S. and Schneider, D. (1982). An analog of ACTH/MSH (4–9), ORG-2766 reduces cerebral uptake of morphine. *Peptides*, **3**, 649–653
159. Fave, G.D., Annibale, B., De Magistris, L., Severi, C., Bruzzone, R., Pouti, M., Melchiorri, P., Torsoli, A. and Erspamer, V. (1985). Bombesin effects on human GI functions. *Peptides*, **6** (Suppl. 3), 113–116
160. Goldstein, A., Tachibana, S., Lowney, L.I., Hunkapiller, M. and Hood, L. (1979). Dynorphin-(1–13), an extraordinarily potent opioid peptide. *Proc. Natl. Acad. Sci. USA*, **76**, 6666–6670
161. Sawyer, T.K., Sanfilippo, P.J., Hruby, V.J., Engel, M.H., Heward, C.B., Burnett, J.B. and Hadley, M.E. (1980). 4-Norleucine, 7-D-phenylalanine-alpha-melanocyte-stimulating hormone: a highly potent alpha-melanotropin with ultralong biological activity. *Proc. Natl. Acad. Sci. USA*, **77**, 5754–5758
162. Burbach, J.P., Kovacs, G.L., de Wied, D., van Nispen, J.W. and Greven, H.M. (1983). A major metabolite of arginine vasopressin in the brain is a highly potent neuropeptide. *Science*, **221**, 1310–1312
163. Pals, D.T., Masucci, F.D. and Denning, G.S. (1971). Role of the pressor action of angiotensin II in experimental hypertension. *Circ. Res*, **29**, 673–181
164. Vavrek, R.J. and Stewart, J.M. (1985). Competitive antagonists of bradykinin. *Peptides*, **6**, 101–164
165. Siggins, G.R. and Groul, D.L. (1986). Synaptic mechanisms in the vertebrate central nervous system. In: *Handbook of Physiology*, ed., F.E. Bloom. Volume on Intrinsic Regulatory Systems of the Brain, The American Physiological Society, Bethesda,Maryland, 1–114
166. Snyder, S.H. (1980). Brain peptides as neurotransmitters. *Science*, **209**, 976–983
167. Bloom, F.E. (1984). The functional significance of neurotransmitter diversity. *Am. J. Physiol.*, **246**, C184–C194
168. Bloom, F.E. (1984). Chemical integrative processes in the central nervous system. In: *Handbook of Chemical Neuroanatomy*, 51–58
169. Bloom, F.E. (1987). Molecular diversity and cellular functions of neuropeptides. In: *Neuropeptides and Brain Function*, eds., E.R. de Kloet, N.M. Wiegant and D. de Wied. *Prog. in Brain Res.*, Vol. 72, 213–223
170. de Wied, D. (1987). The neuropeptide concept. In: *Neuropeptides and Brain Function*, eds., E.R. de Kloet, N.M. Wiegant and D. de Wied, *Prog. Brain Res.*, Vol. 72, 93–108
171. Iversen, L.L., Lee, C.M., Gilbert, R.F., Hunt, S. and Emson, P.C. (1980). Regulation of neuropeptide release. *Proc. R. Soc.*, **210**, 91–111
172. Buijs, R.M. and Van Heerikhuize, J.J. (1982). Vasopressin and oxytocin release in the brain – a synaptic event. *Brain Res.*, **252**, 71–76
173. Lackoff, A. and Jackson, I.M.D. (1981). Calcium dependency of potassium-stimulated thyrotropin-releasing hormone secretion from rat neurohypophysis *in vitro*. *Neurosci. Lett.*, **27**, 17
174. Urban, I.J.A. (1981). Brain vasopressin: from electrophysiological effects to neurophysiological function. In: *Neuropeptides and Brain Function*, eds., E.R. de Kloet, V.M. Wiegant and D. de Wied, *Progress in Brain Research*, Vol. 72, 163–172

175. Dodd, J. and Kelly, J.S. (1981). The actions of cholecystokinin and related peptides on pyramidal neurones of the mammalian hippocampus. *Brain Res.*, **205**, 337–350
176. North, A.R. (1986). Electrophysiological effects of neuropeptides. In: *Neuropeptides in Neurologic and Psychiatric Disease*, eds., J.B. Martin and J.D. Broebes. (Raven Press, New York), 71–77
177. Kelly, J.S. (1982). Electrophysiology of peptides in the central nervous system. *Br. Med. Bull.*, **38**, 283–290
178. North, R.A. and Egan, T.M. (1982). Electrophysiology of peptides in the peripheral nervous system. *Br. Med. Bull.*, **38**, 291–296
179. Hokfelt, T., Lundberg, J.M., Schutzberg, M., Johansson, O., Ljungdahl, A. and Rehfeld, J. (1980). Coexistence of peptides and putative transmitters in neurons. In: *Neural Peptides and Neuronal Communication*, eds., E. Costa and M. Trabucchi, (Raven Press, New York), 1–23
180. Krieger, D.T. (1986). An overview of neuropeptides. In: *Neuropeptides in Neurologic and Psychiatric Disease*, eds., J.B. Martin and J.D. Barkas, (Raven Press, New York), 1–32
181. Quirion, R. (1983). Interactions between neurotensin and dopamine in the brain: an overview. *Peptides*, **4**, 609–615
182. Meisenberg, G. and Simmons, W.H. (1985). Motor hypoactivity induced by neurotensin and related peptides in mice. *Pharmacol. Biochem. Behav.*, **22**, 189–193
183. Kalivas, P.W. (1985). Interactions between neuropeptides and dopamine neurons in the ventromedial mesencephalon. *Neurosci. Biobehav. Rev.*, **9**, 573–587
184. Kalivas, P.W., Burgess, S.K., Nemeroff, C.B. and Prange, Jr., A.J. (1983). Behavioral and neurochemical effects of neurotensin microinjection into the ventral tegmental area of the rat. *Neuroscience*, **8**, 495–505
185. Sarrieau, A., Javoy-Agid, F., Kitabgi, P., Dussaillant, M., Vial, M., Vincent, J.P., Agid, Y. and Rostène, W.H. (1985). Characterisation and autoradiographic distribution of neurotensin binding sites in human brain. *Brain Res.*, **348**, 375–80
186. Rostene, W.H., Sarricau, A., Moyse, E., Hervé, D., Kitalegi, P., Ewen, McB.S., Vial, M., Tassin, J.P., Vincent, J.P. and Beaudet, A. (1987). Imaging of neuropeptides–neurotransmitter interactions. In: *Neuropeptides and Brain Function*, eds., E.R. de Kloet, N.M. Wiegant and D. de Wied, *Prog. Brain Res.*, Vol. 72, 213–223
187. Quiron, R., Chiueh, C.C., Everist, H.D. and Pert, A. (1985). Comparative localization of neurotensin receptors on nigrostriatal and mesolimbic dopaminergic terminals. *Brain Res.*, **327**, 385–389
188. Uhl, G.R. and Kuhar, M.J. (1984). Chronic neuroleptic treatment enhances neurotensin receptor binding in human and rat substantia nigra. *Nature London*, **309**, 350–352
189. Rostene, W.H., Herve, D., Kitabgi, P., Magre, J. and Sarrieau, A. (1986). Hormonal receptor plasticity in the brain as shown by *in vitro* quantitative autoradiography. In: *Neuroendocrine Molecular Biology*, eds., G. Fink, A.J. Harmer and K.W. McKerns, (Plenum Press, New York)
190. Agnati, L.F., Fuxe, K., Battistini, N., Giardino, L., Benfenati, F., Martire, M. and Ruggeri, M. (1985). Further evidence for the existence of interactions between receptors for dopamine and neurotensin. Dopamine reduces the affinity and increases the number of $^{3}$H-neurotensin binding sites in the subcortical limbic forebrain of the rat. *Acta Physiol. Scand.*, **124**, 125–128
191. Agnati, L.F., Fuxe, K., Benfenati, F. and Battistini, N. (1983). Neurotensin in vitro markedly reduces the affinity in subcortical limbic $^{3}$H-*N*-propyl-norapomorphine binding sites. *Acta Physiol. Scand.*, **119**, 459–461
192. Simasko, S.M. and Weiland, G.A. (1985). Effect of neurotensin, substance P and TRH on the regulation of dopamine receptors in rat brain. *Eur. J. Pharmacol.*, **106**, 653–656
193. Magistretti, P.J. and Morrison, J.H. (1985). VIP neurons in the neocortex. *TINS*, **8**, 7–8
194. Morrison, J.H. and Magistretti, P.J. (1983). Monoamines and peptides in cerebral cortex: contrasting principles of cortical organization. *TINS*, **6**, 146–151

195. Magistretti, P.J. and Schorderet, M. (1985). Norepinephrine and histamine potentiate the increases in cyclic adenosine 3′:5′-monophosphate elicited by vasoactive intestinal polypeptide in mouse cerebral cortical slices: mediation by $\alpha_1$-adrenergic and $H_1$-histaminergic receptors. *J. Neurosci.*, **5**, 363–368
196. Ferron, A., Siggins, G.R. and Bloom, F.E. (1985). Vasoactive intestinal polypeptide acts synergistically with noradrenaline to depress spontaneous discharge rates in cerebral cortical neurons. *Proc. Natl. Acad. Sci. USA*, **82**, 8810–8812
197. Rostene, W.H. (1984). Neurobiological and neuroendocrine functions of the vasoactive intestinal peptide (VIP). *Prog. Neurobiol.*, **22**, 103–129
198. Bosler, O. and Beaudet, A. (1985). VIP neurons as prime synaptic targets for serotoninergic afferents in rat suprachiasmatic nucleus: a combined radioimmunological and immunocytochemical study. *J. Neurocytol.*, **14**, 749–763
199. Mancillas, J.R., Siggins, G.R. and Bloom, F.E. (1986). Systemic ethanol: selective enhancement of responses to acetylcholine and somatostatin in hippocampus. *Science*, **231**, 161–163
200. Mancillas, J.R., Siggins, G.R. and Bloom, F.E. (1986). Somatostatin-selectively enhances acetylcholine-induced excitations in rat hippocampus and cortex. *Proc. Natl. Acad. Sci. USA*, **83**, 7518–7521
201. Kordon, C., Blauet-Pajot, M.T., Clausen, H., Drouva, S., Enjabert, A. and Epelbaum, Y. (1987). New designs in neuroendocrine systems. In: *Neuropeptides and Brain Function*, eds., E.R. de Kloet, N.M. Wiegant and D. de Wied, *Prog. in Brain Res.*, **72**, 27–34
202. Michel, D., Lefevre, C. and Labrie, F. (1983). Interactions between growth hormone releasing factor, prostaglandin E2 and somatostatin on cyclic AMP accumulations in rat adenohypophysial cells in culture. *Mol. Cell. Endocrinol.*, **33**, 255–264
203. Robertson, R.P. (1986). Characteristics and regulation of prostaglandin and leucotriene receptors: an overview. *Prostaglandins*, **31**, 395–411
204. Watson, S.P., Connell, R.Mc. and Lapetina, E.G. (1984). The rapid formation of inositol phosphates in human platelets stimulated by thrombin is inhibited by prostacycline. *J. Biol. Chem.*, **259**, 13199–13203
205. Sibley, D. and Lefkowitz, R. (1985). Molecular mechanisms of receptor desensitization using the beta-adrenergic receptor coupled adenylate cyclase system as a model. *Nature (London)*, **317**, 124–129
206. Naor, Z. and Eli, Y. (1985). Synergistic stimulation of luteinizing hormone (LH) release by protein kinase C activator and $Ca^{2+}$ ionophore. *Biochem. Biophys. Res. Commun.*, **130**, 848–853
207. Soper, B.D. and Weick, R.F. (1980). Hypothalamic and extrahypothalamic mediation of pulsatile discharges of luteinizing hormone in the ovariectomized rat. *Endocrinology*, **106**, 348–355
208. Drouva, S.V. and Gallo, R.V. (1976). Catecholamine involvement in episodic luteinizing hormone release in adult overiectomized rats. *Endocrinology*, **99**, 651–656
209. Gudelsky, G.A. and Porter, J.C. (1979). Morphine and opioid peptide induced inhibition of the release of dopamine from tuberoinfundibular neurons. *Life Sci.*, **25**, 1697–1702
210. Drouva, S.V., Epelbaum, J., Tapia-Arancibia, L., Laplante, E. and Kordon, C. (1981). Opiate receptors modulate LHRH and SRIF release from mediobasal hypothalamic neurons. *Neuroendocrinology*, **32**, 163–168
211. Miki, N., Ono, M. and Shizumi, K. (1984). Evidence that opiotergic and alpha-adrenergic mechanisms stimulate rat growth hormone release via growth hormone releasing factor (GRF). *Endocrinology*, **114**, 1950–1952
212. Wehrinberg, W.B., Block, B. and Ling, N. (1985). Pituitary secretion of growth hormone in response to opioid peptide and opiates is mediated through growth hormone releasing factor. *Neuroendocrinology*, **41**, 13–16
213. Plotsky, P.M. and Vale, W. (1985). Patterns of growth hormone releasing factor and somatostatin secretion into the hypohysial portal circulation of the rat. *Science*, **230**, 461–463

214. Drouva, S., Laplante, E. and Kordon, C. (1982). I-Adrenergic receptor involvement in the LH surge in ovariectomized estrogen primed rats. *Eur. J. Pharmacol.*, **81**, 341–344.
215. Honma, K. and Wuttke, W. (1980). Norepinephrine and dopamine turnover rates in the medial preoptic area and the mediobasal hypothalamus of the rat brain after various endocrinological manipulations. *Endocrinology*, **106**, 1848–1853
216. Flugge, G., Oertel, W.H. and Wuttke, W. (1986). Evidence for estrogen-receptive Gaba-ergic neurons in the preoptic/anterior hypothalamic area of the rat brain. *Neuroendocrinology*, **43**, 1–5
217. Rossier, J., French, E., Guillemen, R. and Bloom, F.E. (1980). On the mechanisms of simultaneous release of immunoreactive beta-endorphin and prolactin by stress. *Adv. Biochem. Psychopharmacol.*, **22**, 363–375
218. Selye, H. (1950). *Stress. The Physiology and Pathology of Exposure to Stress*, Acta Medica Publ., Montreal
219. Palkovits, M. (1987). Organization of the stress response at the anatomical level. *Prog. Brain Res.*, **72**, 47–55
220. Bohus, B., Benus, R.F., Fokkema, D.S., Koolheas, J.M., Nyakas, C., van Oortmerssen, G.A., Prins, A.J.A., de Ruiter, A.J.H., Scheurink, A.J.W. and Steffers, A.B. (1987). Neuroendocrine states and behavioral and physiological stress responses. *Prog. Brain Res.*, **72**, 57–70
221. Allen, J.P., Allen, C.F., Greer, M.A. and Jacobs, J.J. (1973). Stress-induced secretion of ACTH. In: *Brain–Pituitary–Adrenal Interrelationships*, ed., A. Brodish. (Karger, Basel), 99–127
222. Ganong, W.F. (1980). Neurotransmitters and pituitary function: regulation of ACTH secretion. *Fed. Proc.*, **39**, 2923–2930
223. Makara, G.B. (1985). Mechanisms by which stressful stimuli activate the pituitary-adrenal system. *Fed. Proc.*, **45**, 149–153
224. Rivier, C. and Vale, W. (1985). Effects of corticotropin-releasing factor, neurohypophyseal peptides, and catecholamines on pituitary function. *Fed. Proc.*, **44**, 189–195
225. Tilders, F.J.H., Berkenbosch, F., Vermes, J., Linton, E.A. and Smelik, P.G. (1985). Role of epinephrine and vasopressin in the control of the pituitary-adrenal response to stress. *Fed. Proc.*, **44**, 155–160
226. Gillies, G.E., Linton, E.A. and Lowry, P.J. (1982). Corticotropin releasing activity of the new CRF is potentiated several times by vasopressin. *Nature (London)*, **229**, 355–357
227. Enjalbert, A., Sladeczek, F., Guillon, G., Bertrand, P., Shu, C., Epelbaum, J., Garcia-Sainz, J.A., Jard, S., Lombard, C., Kordon, C. and Bockaert, J. (1986). Angiotensin II and dopamine modulate both cAMP and inositol phosphate production in anterior pituitary cells. Involvement in prolactin secretion. *J. Biol. Chem.*, **262**, 4071
228. Huang, M. and Rorstad, O.P. (1983). Effects of vasoactive intestinal polypeptide, monoamines, prostaglandins, and 2-chloroadenosine on adenylate cyclase in rat cerebral microvessels. *J. Neurochem.*, **40**, 719–725
229. Huang, M., Hanley, D.A. and Rorstad, O.P. (1983). Parathyroid hormone stimulates adenylate cyclase in rat cerebral microvessels. *Life Sci.*, **32**, 1009–1014
230. Speth, R.C. and Harik, S.I. (1985). Angiotensin II receptor binding sites in brain microvessels. *Proc. Natl. Acad. Sci.*, **82**, 6340–6343
231. Akmal, M.D., Goldstein, A., Multani, S. and Massry, S.G. (1984). Role of uremia, brain calcium and parathyroid hormone on changes in electroencephalogram in chronic renal failure. *Am. J. Physiol.*, **246**, F575–F579
232. Derian, C.K. and Moskowitz, M.A. (1986). Polyphosphoinoside hydrolysis in endothelial cells and carotid artery segments. Bradykinin-2 receptor stimulation is calcium-dependent. *J. Biol. Chem.*, **261**, 3831–3837
233. Hess, J., Gjedde, A. and Jessen, H. (1987). Vasopressin receptors at the blood–brain barrier in rats. *Wiss. Z. Karl-Marx-Univ. Leipzig, Math.-Naturwiss. R.*, **36**, 81–83
234. Spatz, M., Yamamoto, H., Lust, D.W., Wroblewska, B., Merkel, N. and Bembry, J. (1988). Peptides and cerebral microvessels. In: *Peptide and Amino Acids in the Central*

*Nervous Systems*, eds., Lj. Rakic, D.J. Begley, H. Davson and B.V. Zlokovic, (Macmillan, London), 35–43
235. Kretzschmar, R., Landgraf, R., Gjedde, A. and Ermisch, A. (1986). Vasopressin binds to microvessels from rat hippocampus. *Brain Res.*, **380**, 325–330
236. Ermisch, A. (1987). Blood–brain barrier and peptides. *Wiss. Z. Karl-Marx-Univ. Leipzig, Marth.-Naturwiss. R.*, **36**, 72–77
237. Ermisch, A., Landgraf, R., Brust, P., Kretzschmar, R. and Hess, J. (1988). Peptide receptors of the cerebral capillary endothelium and the transport of amino acids across the blood–brain barrier. In: *Peptide and Amino Acid Transport Mechanisms in the Central Nervous System*, eds., Lj. Rakic, D.J. Begley, H. Davson and B.V. Zlokovic, (Macmillan, London), 43–55
238. Petter, H. (1985). Immunozytochemische Untersuchungen zur moglichen Beteiligung des klassischen neurosekretorischen Systems an cerebralen Prozessen der Herz-Kreislauf-Regulation. *J. Hirnforsch.*, **26**, 477–496
239. Reith, J., Ermisch, A. and Gjedde, A. (1987). Saturable vasopressin retention in hippocampal endothelium and in non-BBB brain regions (pineal and pituitary glands). *Wiss. Z. Karl-Marx-Univ. Leipzig, Math.-Naturwiss. R.*, **36**, 87–90
240. Hess, J., Gjedde, A. and Jesscen, H. (1987). Vasopressin receptors at the blood–brain barrier in rats. *Wiss. Z. Karl-Marx-Univ. Leipzig Math.-Naturwiss. R.*, **36**, 81–83
241. Zlokovic, B.V., Begley, D.J., Segal, M.B., Davson, H., Rakic, Lj., Lipovac, N., Mitrovic, D.M. and Jankov, R.M. (1988). Neuropeptide transport mechanisms in the central nervous system. In: *Peptide and Amino Acid Transport in the Central Nervous System*, eds., Lj. Rakic, D.J. Begley, H. Davson and B.V. Zlokovic, (Macmillan, London), 3–21
242. Ermisch, A., Landgraf, R. and Mobius, P. (1986). Vasopressin and oxytocin in brain areas of rats with high or low behavioral performance. *Brain Res.*, **379**, 21–29
243. Pardridge, W. (1986). Receptor-mediated peptide transport through the blood–brain barrier. *Endocr. Rev.*, **7** (3), 314–33
244. Davson, H., Welch, K. and Segal, M.B. (1987). *Physiology and Pathophysiology of the Cerebrospinal Fluid* (Churchill Livingstone, Edinburgh)
245. van Houten, M. and Posner, B.I. (1983). Circumventricular organs: receptors and mediators of direct peptide hormone action on brain. In: *Adv. Metab. Disorders Vol. 10*, ed., A. Szabo, (Academic Press, New York), 269–289
246. Baskin, D.G., Woods, S.C., West, D.B., van Houten, M., Posner, B.I., Dorsa, D.M. and Porte Jr., D. (1983). Immunocytochemical detection of insulin in rat hypothalamus and its possible uptake from cerebrospinal fluid. *Endocrinology*, **112**, 1818–1825
247. Baskin, D.G., Dorsa, D.M., Figlewicz, D.P., Corp, E.S., Wilcox, B.J., Wallum, B.J. and Wood, S.C. (1988). Insulin as a regulatory peptide. In: *Peptide and Amino Acid Transport Mechanisms in the Central Nervous System*, eds., Lj. Rakic, D.J. Begley, H. Davson and B.V. Zlokovic, (Macmillan, London), 81–93
248. McKinley, M.J., Allen, A., Clevers, T., Dentin, D.A., Mendlesohn, F.A.O., Oldfield, B.J., Tarjan, E. and Weisirger, R.S. (1987). Angiotensin II receptors in the brain of the sheep. *Wiss. Z. Karl-Marx-Univ. Leipzig Math.-Naturwiss. R.*, **36**, 189–192
249. Walsh, R.J., Slaby, F. and Posner, B.I. (1987). Prolactin transport from blood to cerebrospinal fluid: a receptor mediated process. *Wiss. Z. Karl-Marx-Univ. Leipzig Math.-Naturwiss. R.*, **36** (1), 119–120
250. Uddman, R., Edvinsson, L., Owman, C. and Sundler, F. (1981). Perivascular substance P: occurrence and distribution in mammalian pial vessels. *J. Cereb. Blood Flow Metab.*, **1**, 227–232
251. Kapadia, S.E. and DeLanerolle, N.C. (1984). Immunohistochemical and electron microscopic demonstration of vascular innervation in the mammalian brainstem. *Brain Res.*, **292**, 33–39
252. Uddman, R., Edvinsson, L., Owman, C. and Sundler, F. (1983). Nerve fibres containing gastrin-releasing peptide around pial vessels. *J. Cereb. Blood Flow Metab.*, **3**, 386–390

253. Pardridge, W.M. (1986). Blood–brain barrier: interface between internal medicine and the brain. *Ann. Intern. Med.*, **105**, 82–95
254. Chan-Palay, V. (1977). Innervation of cerebral blood vessels by norepinephrine, indoleamine, substance P and neurotensin fibres and the leptomeningal indoleamine axons: their role in vasomotor activity and local alterations of brain blood composition. In: *Neurogenic Control of the Brain Circulation*, eds., C. Owman and L. Edvinsson, (Pergamon Press, Oxford), 39–46
255. Owman, C., Hanko, J., Hardebo, J.E. and Kahrstrom, J. (1986). Neuropeptides and classical autonomic transmitters in the cardiovascular system: existence, coexistence, action, interaction. In: *Neural Regulation of Brain Circulation*, eds., C. Owman and J.E. Hardebo, (Elsevier Science Publishers, B.V. (Biomedical division) Amsterdam), 299–331
256. Hanko, J., Hardebo, J.E. and Owman, C. (1981). In: *Cerebral Microcirculation and Metabolism*, eds., J. Cervos-Navarro and E. Fritschka, (Raven Press, New York), 157–161
257. Banks, W.A. and Kastin, A.J. (1988). Peptides and the blood–brain barrier. In: *Peptides and Amino Acid Transport Mechanisms in the Central Nervous System*, eds., Lj. Rakic, D.J. Begley, H. Davson and B.V. Zlokovic, (Macmillan, London), 21–32
258. Edvinsson, L., Fahrenburg, J., Hanko, J., McCulloch, J., Owman, C. and Uddman, R. (1981). Vasoactive intestinal polypeptide and effects on cerebral blood flow and metabolism. In: *Cerebral Blood Flow and Metabolism*, eds., J. Cervos-Navarro and E. Fritschka, (Raven Press, New York), 147–155
259. McCulloch, J. and Edvinsson, L. (1980). Cerebral circulatory and metabolic effects of vasoactive intestinal polypeptide. *Am. J. Physiol.*, **238**, H449–H456
260. Haffman, S.J., Heilgers, J., Dzoljich, R. and Saxera, P.R. (1986). Regional cerebral blood flow during enkephalin-induced seizures in the rat. *Neuropharmacology*, **25**, 361–365
261. Davson, H. and Segal, M.B. (1970). The effects of some inhibitors and accelerators of sodium transport on the turnover of $^{22}$Na in the cerebrospinal fluid. *J. Physiol.*, **209**, 131–153
262. Rap, Z.M., Kozniewska, E. and Skolasinska, K. (1980). Effect of vasopressin on cerebral blood flow and cerebrospinal fluid outflow. In: *Pathophysiology and Pharmacotherapy of Cerebrovascular Disorders*, eds., E. Betz, J. Grote, D. Hauser, (Verlag G. Witzstrock, Köln), 12–14
263. Rap, Z.M. (1981). Inhibitory effect of antidiuretic hormone on outflow of the cerebrospinal fluid in vasogenic brain edema induced by cold lesion. In: *Cerebral Microcirculation and Metabolism*, eds., J. Cervos-Navarro and E. Fritschke, (Raven Press, New York), 171–175
264. Lindvale, M., Alumets, J., Edvinsson, L., Fahrenkrug, J., Hakanson, R., Hanko, J., Owman, C., Schaffalitzky de Muckadell, O.B. and Sundler, F. (1978). Peptidergic (VIP) nerves in the mammalian choroid plexus. *Neurosci. Lett.*, **9**, 77–82
265. Zlokovic, B.V., Segal, M.B., Davson, H. and Jankov, R.M. (1988). Passage of delta-sleep-inducing peptide (DSIP) across the blood–cerebrospinal fluid barrier. *Peptides*, **9**, 533–538
266. Landgraf, R., Hess, J. and Hartmann, E. (1977). Der Einfluss von Ocytocin auf die regionale $^{3}$H Orotsaure-Aufnahme durch das Rattengehirn. *Endokrinologie*, **70**, 45–52
267. Landgraf, R., Hess, J. and Ermisch, A. (1978). The influence of vasopressin on the regional uptake of [$^{3}$H]orotic acid by rat brain. *Acta Biol. Med. Ger.*, **37**, 655–658
268. Ermisch, A., Landgraf, R. and Neumeister, D. (1980). Blood–brain barrier permeability and vasopressin. *Proc. Int. Union Physiol. Sci.*, **14**, 399
269. Brust, P. and Ermisch, A. (1984). Alteration of 1-[$^{3}$H]leucine transport across the blood–brain barrier (BBB) elicited by arginine vasopressin (AVP). In: *Regulation of Transmitter Function: Basic and Clinical Aspects. Proc. Fifth Meeting Eur. Soc. Neurochem.*, eds., K. Magyar and E.S. Vizi, Budapest, Akademiai Kiado, 47

270. Raichle, M.E. and Grubb, R.I. (1978). Regulation of brain water permeability by centrally-released vasopressin. *Brain Res.*, **143**, 191–194
271. Raichle, M.E. (1981). Hypothesis: a central neuroendocrine system regulates brain ion homeostasis and volume. In: *Neurosecretion and Brain Peptides*, eds., J.B. Martin, S. Reichlin and K.L. Bick, (Raven Press, New York), 329–336
272. Barry, D.I., Paulson, O.B. and Hertz, M.M. (1980). The blood–brain barrier: an overview with special reference to insulin effect on glucose transport. *Acta Neurol. Scand.*, **778**, 147–156
273. Hertz, M.M., Paulson, O.B., Barry, D.I., Christiansen, J.S. and Svendsen, P.A. (1981). Insulin increases glucose transfer across the blood–brain barrier in man. *J. Clin. Invest.*, **67**, 597–604
274. Goldman, H. and Murphy, S. (1981). An analog of ACTH/MSH ORG-2766, reduces permeability of the blood–brain barrier. *Pharmacol. Biochem. Behav.*, **14**, 845–848
275. Crone, C. (1986). The blood–brain barrier: a modified tight epithelium. In: *The Blood–Brain Barrier in Health and Disease*, eds., A.J. Suckling, M.G. Rumsby and M.W.B. Bradbury, (Ellis Harwood, Chichester, England), 17–40
276. Joó, F. (1986). New aspects to the function of the cerebral endothelium. *Nature (London)*, **321**, 197–198
277. Joó, F. (1988). Cyclic nucleotide-mediated regulation of albumin transport in brain microvessels. In: *Peptide and Amino Acid Transport in the Central Nervous System*, eds., Lj. Rakic, D.J. Begley, H. Davson and B.V. Zlokovic, (Macmillan), 121–131
278. Pardridge, W.M., Yang, J. and Eisenberg, J. (1985). Blood–brain barrier protein phosphorylation and dephosphorylation. *J. Neurochem.*, **45**, 1141–1147
279. Banks, W.A. and Kastin, A.J. (1986). Modulation of the carrier mediated transport of Tyr-MIF-1 across the blood–brain barrier by essential amino acids. *J. Pharm. Exp. Therap.*, **239**, 668–672
280. Stewart, P.A. and Wiley, M.J. (1981). Developing nervous tissue induces formation of blood–brain barrier characteristics in invading endothelial cells: a study using quail-chick transplantation chimeras. *Dev. Biol.*, **84**, 183–192
281. Pardridge, W.M., Yang, J., Eisenberg, J. and Mietus, L.J. (1986). Antibodies to blood–brain barrier bind selectively to brain capillary endothelial lateral membranes and to 46K protein. *J. Cereb. Blood Flow Metab.*, **6**, 203–211
282. Pardridge, W.M. (1986). Mechanisms of neuropeptide interaction with the blood–brain barrier. *Ann. New York Acad. Sci.*, **481**, 231–249

# 5 Brain peptides and peptidergic pathways

A number of peptides have been characterized and recognized as members of a family of *brain peptides* or *neuropeptides*. These are by definition peptides which have been shown to exist within the cells, axons and terminals of the central nervous system. Neuropeptides have been localized in different brain areas, and peptidergic pathways have been visualized in the central nervous system, by means of various immunocytochemical, immunofluorescence and radioreceptor assay studies. Most neuropeptides have been isolated from brain tissue and their chemical structure identified. During the early 1950s, for some neuropeptides, the regulation of the *hypothalamo–hypophysiotropic axis* was postulated. However, it was not until 1969, when the new era in *neuroendocrinology* was started with the isolation of TRH from ovine and porcine hypothalamic tissue and identification of its chemical structure as a tripeptide-amide (L-pyroglutamyl-L-histidyl-L-proline amide), that the authenticity of the Harris' portal-vessel chemotransmitter hypothesis gained firm experimental evidence. The availability of TRH in a pure form led to the development of the specific radioimmunoassays for this peptide, and subsequently a few years later in 1974, TRH was localized in the extrahypothalamic brain regions, as well as outside the central nervous system. Shortly after the discovery and identification of TRH in 1974 two other hypophysiotropic hormones, LHRH and SRIF, were isolated and demonstrated in the extrahypothalamic brain areas, while a few years later in 1981 and 1982 the location of CRF and GHRH in the extrahypothalamic regions was described.

Although hypothalamic–hypophysiotropic hormones are transported to the adenohypophysis via the portal vessel circulation, as classical hormones, to cause the release of TSH and ACTH, and the inhibition or release of GH, there is much evidence to suggest that these peptides, in their extrahypothalamic loci in the CNS, play a role in neurotransmission and/or neuromodulation, and in cell-to-cell regulation. Thus, apart from their *endocrine* functions, peptides also have *neurocrine*, *paracrine* and *autocrine* functions (Table 5.1). So the same peptide can act in some cases as a local factor, neurotransmitter, or at distance as a hormone. In 1978, in addition to hypothalamic–hypophysiotropic hormones, three nonapeptides all closely related by structure, i.e. the *neurohypophyseal hormones*, oxytocin and VP(Arg) and the *pineal gland hormone*, VT(Arg) were found to be distributed within the central nervous system and released at extrahypothalamic, extrapituitary and extrapineal nerve terminals. Although the systemic endocrine functions of oxytocin are concerned with uterine contraction and with lactation, and that of VP(Arg) with water balance and vasopressor effects, these nonapeptides, when released at other sites in the central nervous system have different functions.

**Table 5.1** Peptide functions classified with respect to the mode by which a secreted peptide may reach its target cell (according to Kreiger (1986)[1])

| *Peptide function* | *Mode of cell-to-cell communication* | *Secreting tissue* | *Peptide action* |
|---|---|---|---|
| Endocrine | At a distance | Gastrointestinal tract<br>Hypophysis<br>Pineal gland | Hormonal regulation of distant tissue functions |
| Paracrine | Local | Gastrointestinal tract<br>Hypophysis | Regulation of the functions of other cells in the same tissue from which peptide is secreted |
| Autocrine | Local | Tissue cell<br>Unicellular organism | Autoregulation of the tissue cell and unicellular organism functions |
| Neurocrine | Local | Central nervous system<br>Ganglia | Neurotransmitter and/or neuromodulator synaptic actions on nerve cells |
| | | Sympathoadrenal tissue | Neuroendocrine regulation of hormonal secretion from the adrenal medulla |

Amongst the peptides, one group until recently had a less controversial history[2]. In 1902, Bayliss and Starling[3] described secretin as the first *gastrointestinal (GI) hormone*, and later other hormones of this group, such as gastrin and CCK, were found. For a long period it was understood that these GI hormones were only concerned with the regulation of digestion. This was in contrast to the situation in neuroendocrinology where it had been accepted that the actions of the adenohypophyseal hormones were mediated only through the secondary control of the hypothalamic–hypophysiotropic peptides.

**Table 5.2** Categories of mammalian neuropeptide families classified according to the original tissue from where peptide was discovered or to some functional aspects (modified according to Kreiger (1986)[1])

**GI-peptides**
- BBS
- CCK
- Gastrin
- Glucagon
- Enk(Leu)
- Enk(Met)
- Insulin
- Motilin
- NT
- PP
- Secretin
- SP
- SRIF
- TRH
- VIP

**Growth factors**
- EGF
- Endothelial cell growth factor
- FGF
- IGFII
- NGF

**Hypophyseal peptide**
- ACTH
- GH
- LH
- β-LPH
- α-MSH
- PRL
- TSH

**Hypothalamic–hypophysiotropic hormones**
- CRF
- GHRH
- LHRH
- SRIF
- TRH

**Neurohypophyseal hormones**
- Neurohypophysin(s)
- OT
- VP(Arg)

**Opioid peptides**
- Dynorphin
- β-E
- Enk(Leu)
- Enk(Met)

**Others**
- ANF
- Angiotensin II
- BBS
- Bradykinin
- Carnosin
- CGRP
- CT
- DSIP
- NY
- Thymosin

GI hormone secreting cells in the gut and brain have the same embryological origin. The first finding to link GI peptide physiology to some more general mechanism was the concept of APUD (amine precursor uptake and decarboxylation) system proposed by Pearse[4,5]. However, even more striking was the demonstration of gastrin-like immunoreactivity in the brain in 1975[6] followed shortly by similar findings for other GI peptides, such as VIP and CCK in 1976, and insulin and glucagon in 1978 and 1979, respectively. New peptides within the so called gut–brain peptide family were subsequently discovered with some of its members also being found in the skin, lung and other tissues[7,8]. Different categories of mammalian brain peptides, classified with respect to the original tissues from where they were first discovered, or according to some functional aspects, are given in Table 5.2.

The primary structure for a given peptide is often the same among its various tissues, as for example NT, SRIF or substance P are identical in the gut and the brain, although the predominant form may vary, as for instance CCK-8 in the brain and CCK-33 in the gut[9]. However, with regard to the nature of receptors in various tissues for the same peptide, as for example insulin, there is evidence that receptors present in the brain may differ from those in the GI tract[10]. In this Chapter we will be primarily interested in the anatomical distribution of various peptides in the brain, along with a consideration of their functional significance in the central nervous system of humans and other species, bearing in mind the concept of *multiple independent actions* of neuropeptides formulated by Kastin and co-workers[11,12].

## ANATOMICAL DISTRIBUTION OF PEPTIDES IN THE CENTRAL NERVOUS SYSTEM

For most neuropeptides, the localization of their neuronal pathways has been one of the most powerful tools in the understanding of their functions. Anatomical distribution of peptides in the brain has been studied using *qualitative*, *semi-quantitative* and *quantitative* methods. Qualitative and semi-quantitative studies are usually based on either *immunofluorescence*, or light microscopy and *immunocytochemical* methods: the most commonly employed methods being the immunoperoxidase reaction for the identification of peptide localization, or light microscopic *autoradiography* with the receptor labelling technique. Quantitative studies were based either on *radioimmunoassay* (RIA) procedures and the measurement of the peptide concentration in the brain homogenates and CSF, or in the microdissected brain nuclei. Alternatively, extracts from different brain areas may be subjected to *high performance liquid chromatography* (HPLC) following purification on affinity chromatography, or *quantitative autoradiography* (QAR) with computer digital image analysis may be employed to give the concentration of bound peptide in small anatomically defined regions of the brain.

One has to be very careful in making assertions about peptide localization by immunocytochemistry, since the stain produced in a tissue may represent a non-specific reaction to other substances and not the specific peptide to which the antibody was raised. The only test of specificity in such immunohistochemical studies is to estimate whether the staining vanishes when the antibody is absorbed with the peptide being stained, although this control experiment in fact does not rule out the possibility of staining the cross-reacting substance. A cross-reacting substance is usually either a peptide metabolite or its precursor, or some peptide-like substance. To avoid this kind of problem in the visualization of the peptidergic pathways and their brain locations, more reliable results can be obtained by a combination of different methods. For example, in RIA studies one can fractionate brain tissue extracts by procedures with high resolving capacities, such as HPLC, to determine whether the material which interacts with the antibody possesses the same physical characteristics as the authentic peptide. Also a detailed regional quantitative distribution of brain peptides with the appropriate chemical controls may be compared with immunohistochemical or immunofluorescence localization to conclude whether most, if not all of the immunohistochemical data, represent staining of the authentic peptide.

## Enkephalins

In 1975 Hughes *et al.*[13] isolated and determined the structure of two naturally occurring brain *opioid peptides*, methionine-enkephalin Enk(Met) and leucine enkephalin Enk(Leu), and a year later Simantov and Synder[14], using both different assays and isolation procedures and in a different species, isolated and identified the same two pentapeptides. The discovery of opioid brain peptides was closely linked to the identification of *opiate receptors* by molecular binding techniques. Thus it was demonstrated, even before the isolation of enkephalins, that brain extracts contained a substance that may compete for opiate receptor binding, and which had a similar regional brain distribution to opiate receptors. Not only enkephalins were found as endogenous ligands for the opiate receptors, but the subsequent studies revealed an intimate co-incidence between immunohistochemical distribution of enkephalins in the central nervous system and the localization of opiate receptors demonstrated by autoradiography (for review see ref.15). For example, both opiate receptors and enkephalin immunoreactivity were concentrated in a narrow band in substantia gelatinosa of the spinal cord, which represents actually the first way-centre for integrated information for the *pain* in the central nervous system. A similar high density of both opiate receptors and enkephalin immunoactivity was found in other brain areas associated with pain perception, such as the medial thalamus, periaqueductal grey matter, substantia gelatinosa and the locus ceruleus, which is the major norepinephrine containing nucleus in the brain. Such associations of the opiate receptor distribution and enkephalins had led to the suggestion that enkephalins may play an opiate role in the normal functions of the brain, which has been supported by their well-established analgesic properties[1].

Enkephalin-containing pathways in the central nervous system have been visualized by combining some lesion experiments and employing compounds which block axonal flow, such as colchicine. This has permitted nerve cell bodies containing enkephalins to be identified, together with peptidergic nerve terminals and axons. It has been demonstrated that enkephalins are localized in the small interneurons in the substantia gelatinosa in the spinal cord and in several long pathways (for review see refs. 1,15). Localization of enkephalins in small interneurons in the layers I and II of the substantia gelatinosa is different from substance P, which occurs only in the primary sensory neuron terminals in substantia gelatinosa, but not in the cells. Substance P-containing cells are observed in the dorsal root ganglia which do not contain the *enkephalin cells*. It has been established that enkephalin terminals may form axo-dendritic and/or axo-somatic contacts with other neurons such as gamma-aminobutyric acid (GABA) cells, which in turn contact substance P terminals, suggesting a possible indirect interaction between the two peptides in the first integrative way-station for pain[16].

The interrelationships of enkephalins and other peptides involved in pain perception are extremely complex and take place at different levels in the central nervous system (Table 5.3). It has been suggested that enkephalins may inhibit dorsal horn neurons and also block the release of substance P from sensory terminals, as well as modify the release and action of SRIF, VIP, CCK, angiotensin, NT, BBS and pancreatic polypeptide at various areas in the brain involved in the transmission of nociceptive information[1].

**Table 5.3** Enkephalins and other neurotransmitter and/or neuro-modulator peptide candidates involved in the integration of pain perception at different levels of the central nervous system (modified according to Goodman *et al.* (1983)[15] and Kreiger (1986)[1]

| | *Neurotransmitter and/or neuromodulator peptide candidates* | |
|---|---|---|
| **Supraspinal level** | | |
| Periaqueductal grey | SP | OT/VP(Arg) |
| Nucleus raphe magnus | TRH | Angiotensin |
| | Enk | |
| **Spinal level** | | |
| Dorsal horn (Laminar arrangement) | Enk | NT |
| (first integrative way-station) | SP | BBS |
| | SRIF | PP |
| Primary sensory neuron | SP | CCK |
| transmitters | SRIF | Angiotensin |
| | VIP | |

Amongst several enkephalin pathways in the brain, the globus pallidus contains probably the highest concentration of enkephalins; however, peptidergic terminals have been observed only after colchicine treatment. It has been found that enkephalin neurons are actually situated in the caudate nucleus, providing a diffuse input into the globus pallidus. The most prominent and long enkephalin pathway is located within the limbic system with cell bodies in the central nucleus of amygdala, and terminals in the bed nucleus of the stria terminalis and possibly in other zones of the hypothalamus. Another important enkephalin pathway starts with enkephalin cells in the supraoptic and paraventricular nuclei of the hypothalamus, which provide terminals to the posterior pituitary. Besides the above-mentioned *enkephalinergic pathways*, it is noteworthy that enkephalin-containing cell groups are widely spread within the central nervous system; these include the following regions; the perifornical area of the hypothalamus, the interpendicular nucleus, the medial and spinal vestibular nuclei, the dorsal cochlear nucleus, the nucleus of the solitary tract, the Golgi II type cells in the cerebellum, the olfactory bulb and preoptic nucleus, the pyramidal areas of the hippocampus, the neocortex, etc.

Since the discovery of stereospecific opioid binding sites and their endogenous opioid peptides, the hypothesis of receptor multiplicity has become accepted on the basis of a wide range of evidence[17], which involves the existence of several subtypes: $\mu$ (morphine), $\delta$ (enkephalin), $\kappa$ (ethylketocyclazozine) and $\sigma$ (SKF 10047). Despite a great deal of neuropharmacological and biochemical studies, very little is as yet known about their molecular basis, and the principal question is whether they are due to distinct sites at the molecular level, or are multiple states of a single receptor[18].

It was found that enkephalins have affinity predominantly to the $\mu$ and $\delta$ forms of opiate receptors. On the other hand, it has been accepted, on the basis of more specific immunohistochemical studies, that Enk(Met) and Enk(Leu) may be stored in distinct cell populations in the brain, which raises the question as to whether anatomical distinction and distribution of different subtypes of the opiate receptors may have functional correlations with distribution of Enk(Met) and Enk(Leu) neurons. It has been shown that distribution of $\mu$ receptors in general parallels that of Enk(Met) neurons, while $\delta$ receptors fit more with Enk(Leu) localization (Table 5.4). These finding have led to the proposal that since $\mu$ receptors appear to be more concentrated in areas associated with the integration of *pain* information, they may serve as the physiological receptor sites for synaptic action of Enk(Met), whereas $\delta$ receptors are distributed predominantly in the limbic system and may act as the receptor sites for Enk(Leu)[15].

Although at first sight the hypothesis of selectivity of opiate receptors and distinct enkephalin neurons is attractive, it may be more complex than it was originally postulated. Molecular binding experiments in rat brain membranes, with the aid of specific displacing ligands, provided evidence that suggests that the distinct subtypes of the opioid receptors ($\delta$, $\kappa$, $\mu_1$ and $\mu_2$), may undergo allosteric interactions with the formation of heterologous $\kappa$–$\delta$ and $\mu_1$–$\mu_2$ receptor complexes, which is required for stabilization of the high affinity conformational state of the receptor[18,19]. Also, the cations $Mg^{2+}$ and $Na^+$, as well as the guanine nucleotides, may play a role in regulating the equilibria of receptor association–dissociation complexes which may in turn influence their affinity towards the ligands. Hill plots of the uptake by a typical agonist for $\delta$ receptors, DADLE, indicated a higher affinity to $\mu$ receptors than had previously been thought, and the estimated affinity of naloxone (a typical $\mu$ receptor antagonist) for $\delta$ receptors was also significantly higher than previously reported.

**Table 5.4** Distribution of $\mu$ and $\delta$ opiate receptors and corresponding localization of Enk(Met) and Enk(Leu) neurons in the central nervous system (modified from Goodman *et al.* (1983)[15] ( > denotes more; < denotes less; ≈ approximately equal)

| | Opiate receptors | | Encephalinergic neurons | |
|---|---|---|---|---|
| Brain areas | $\mu$ | $\delta$ | Enk(Met) | Enk(Leu) |
| Hippocampus | > | | > | |
| Paraventricular nucleus | > | | > | |
| Periaqueductal grey | > | | > | |
| Thalamus | > | | > | |
| Cerebral cortex | ≈ | | ≈ | |
| Caudate/putamen | ≈ | | ≈ | |
| Globus pallidus | ≈ | | ≈ | |
| Spinal cord | ≈ | | ≈ | |
| Central amygdaloid nucleus | < | | < | |

The above findings have stimulated the development of a number of synthetic peptide analogues for naturally occurring enkephalins, which discriminate between morphine selective ($\mu$) receptors and $\delta$ receptors more effectively than DADLE, both in radioreceptor and in bioassay studies. This group includes the highly selective peptides for $\delta$ receptors such as: D-serine$^{2}$-leucine$^{5}$-enkephalin-threonin$^{6}$ (DSLET), D-pen$^{2}$-L-pen$^{5}$-enkephalin (DPLPE) and D-pen$^{2}$-L-cys$^{5}$-enkephalin (DPLCE). Binding studies have confirmed the high selectivity of DSLET to $\delta$, as opposed to morphine-preferring ($\mu$, ) sites in the brain[20].

The issue of opioid peptides and their physiological role in the central nervous system became even more complex with the discovery of a non-peptide morphine-like compound in normal mammalian tissues, which has been finally identified as *morphine* itself[21]. Furthermore, it was possible to corroborate and characterize morphine from beef brain and adrenal glands[22], and it has also been demonstrated that mammalian tissues can synthesize both morphine and codeine[23], which were found also as endogenous components in human CSF[24]. The results obtained from the chromatographic behaviour and cross-reactivity with both monoclonal and polyclonal antibodies, have indicated the presence of the *opiate alkaloids*, morphine and codeine in human CSF in concentrations comparable to those of the opioid peptides. This may suggest that the normal *opiate* functions of the brain are governed jointly by endogenous opiate alkaloids and opioid peptides. An interesting question, which still seems to be unanswered satisfactorily from the molecular biology point of view, is the mode of action of both alkaloid opiates and opioid peptides, with respect to their interaction with the recognition sites of the opioid receptors. With regard to their different chemical structure and opiate receptor multiplicity, the factors that regulate the state of the receptor recognition site and subsequent translation of the alkaloid and/or peptide binding into the biochemical events, which ultimately lead to a biological response, may probably give a clue for the intimate mechanism of the *opiate function* of the brain.

## Thyrotropin releasing hormone (TRH)

It has been established on the basis of quantitative studies that over 70% of total TRH in brain is found outside the hypothalamus, the majority of which is distributed in the thalamus, brain stem, olfactory lobe and spinal cord. Extrahypothalamic brain tissue from normal human adults (killed in traffic accidents) also contains TRH especially in the thalamus and cerebral cortex[25]. TRH-staining nerve terminals have been demonstrated in several motor nuclei of the brain stem and in the intermediolateral column and terminals around motoneurons of the spinal cord. Immunocytochemical studies have shown that TRH-like and substance P-like peptides occur together in cell bodies of serotonin neurons in the medulla oblongata, which suggests the co-existence of neuropeptides and monoamine in the same nerve endings, as well as their origin from the higher centres[26]. The association of TRH, substance P and serotonin in the same nerve cell strongly argues against the so-called Dale's principle of 'one neuron–one transmitter' (see Table 4.7) and similar co-existence was shown to be widespread for other brain peptides and neurotransmitters. Within the medulla oblongata, TRH has been localized in the nucleus tractus solitarius, nucleus ambigus, the dorsal motor nucleus of the vagus and in neurons innervating the phrenic motor nucleus[27]. Such localization of TRH has suggested that this peptide may take a part as a neurotransmitter and/or neuromodulator in the central nervous regulation of breathing. When the extracts of the hypothalamus, extrahypothalamic brain and spinal cord were analysed by HPLC, following purification by affinity chromatography, the results indicated that tissue TRH and synthetic TRH were identical as well as having similar biological activity. These findings support the view that immunoreactive TRH, (IR)-TRH, found in the brain is the authentic tripeptide.

## Luteinizing hormone-releasing hormone (LHRH)

The decapeptide LHRH was first isolated from porcine and ovine hypothalamic tissue and, following the development of radioimmunoassay, extracts of stalk median eminence of the rat, man, sheep, monkey and pig were shown to cross-react with antisera to LHRH[28]. LHRH releases both FSH and LH from the adenohypophysis and some authors now have renamed this decapeptide as gonadotropin-releasing hormone (GnRH). By using specific anti LHRH antibodies and immunohistochemical techniques, the distribution of LHRH has been demonstrated both in the hypothalamus and extrahypothalamic brain areas, which suggests that LHRH may participate in the synaptic events in the central nervous system independently of its role in regulating adenohypophyseal secretory activity[29]. Immunoreactive (IR)LHRH is also found in numerous ependymal tanacytes, whose function has been postulated to carry hormones released from hypothalamic nerve endings into the CSF, and then to the target cells in different special regions in the brain[30]. This supports the view that the CSF and ventricular system may provide a pathway for carrying LHRH and possibly some other neuropeptides secreted from hypothalamus to their target tissue, i.e. the postulated 'third circulation' within the brain, where they can exert neurohormonal control of specific brain functions, such as feeding and drinking.

Immunohistochemical, as well as RIA studies, have unequivocally revealed the presence of significant amounts of LHRH in the organum vasculosum of the lamina terminalis (OVLT), which is the circumventricular organ in the anterior wall of the third cerebral ventricle with a fine structure similar to that of the median eminence. It has been suggested that OVLT may act as a neurohumoral outlet for hypothalamic peptides into the bloodstream[31], probably mediating transport of LHRH and SRIF, thus taking a part in the control of pituitary function. Mechanical lesions of the OVLT result in an irregular estrous cycle in the rat, reduced mediobasal hypothalamic LHRH content as well as reduced LH and FSH serum levels[32]. These observations support the view that the OVLT plays an important role in the control of cyclic, but not in the tonic, release of the secretion of pituitary gonadotropins. By using local application of cainic acid into the medial preoptic area to induce hyperactivity in the LHRH hyperkarya, evidence was obtained that a separate FSH-RF may exist in the OVLT[32]. By means of the *in situ* hybridization method, using radiolabelled DNA, which through complementary base-pairing, hybridizes to specific sequence of LHRH mRNA forming DNA : RNA hybrids, cells in the rat forebrain which contain LHRH mRNA have been localized[33]. It has been noted that mRNA for LHRH is localized in the rat forebrain in the same regions previously reported to contain immunoreactive LHRH.

## Corticotropin releasing factor

The sequence of ovine corticotropin-releasing factor (CRF) as a 41-amino acid hypothalamic peptide, which is the major physiological stimulus for POMC-derived peptide secretion, has been recently elucidated[34,35]. CRF like anterior hypothalamic regulatory peptide is widely distributed in the central nervous system and is involved in coordination of a variety of endocrine, autonomic and behavioural functions[36]. CRF receptors in the pituitary and brain have been characterized and localized by an *in vitro* labelling autoradiographic procedure using $Nle^{21}$, $^{125}I$-$Tyr^{32}$-oCRF both in the rat and human[36]. It has been demonstrated that the distribution of CRF receptors is similar in the brain areas of the rat, monkey and human (Table 5.5). The distribution of CRF receptors in rat forebrain is consistent with immunohistochemical mapping[37–40] as well as with the proposed sites of its action[41–43]. For instance, CRF cell bodies in the neocortex are concentrated in laminae II and III with projections to laminae I and IV that are rich in CRF receptors. Iontophoretic application of CRF to the membrane of individual neurons produces changes in spontaneous action potentials and the neurons in the cortex and hypothalamus become excited whereas those in the thalamus and lateral septal become inhibited[44]. The presence of CRF receptors in the amygdala, thalamus and septum support these findings. The possibility of identification and localization of CRF receptors in postmortem human brain tissue has provided the basis for elucidating the role of this peptide in the ethology and pathophysiology of a variety of endocrine as well as neurological and psychiatric disorders, such as depression, alcoholism, anorexia nervosa, bulimia, obsessive–compulsive neurosis and dementia[45].

## Somatostatin

The hypothalamic tetradecapeptide, somatostatin (SRIF), inhibits the release of growth hormone by the anterior pituitary, and takes part in the regulation of the secretion of the two pancreatic hormones, insulin and glucagon. Following the development of the appropriate radioimmunoassay, it was shown that high concentrations of SRIF are present in the median eminence and arcuate nucleus of hypothalamus. Immunohistochemical studies, including electron microscopy, have demonstrated SRIF located in the nerve endings of the external layers of the median eminence and in the ventromedial nucleus[30]. The cellular source of SRIF in the medial basal hypothalamus seems to be from SRIF-containing perikarya in the preoptic and anterior pariventricular nuclei, which was shown by immunocytochemical staining.

Although originally isolated from hypothalamic tissue, SRIF was later found in the gastrointestinal tract, the D-cells of pancreas and then found to be widely distributed within the central nervous system, including loci in the cortex, thalamus, striatum, medulla, pons and cerebellum[46,47]. The distribution of endogenous immunoreactive SRIF and the binding affinity of [$^{125}$I]-SRIF in a preparation of membranes from the rat brain synaptosomes from different brain regions are shown in Table 5.6. According to this table, the density of the receptor specific SRIF binding sites and the distribution of endogenous SRIF are different. It is postulated that SRIF participates in the regulation of feeding behaviour, extrapyramidal motor activity, turnover of brain acetylcholine and the release of brain serotonin[30].

**Table 5.5** Distribution of CRF binding sites in rat forebrain. According to de Sousa and Kuhar (1986)[36]

| *Region* | *Relative density of CRF receptors* |
|---|---|
| Frontoparietal cortex (somatosensory) | |
| Laminae I and II | 3+ |
| Laminae III and IV | 4+ |
| Laminae V and VI | 3+ |
| Frontoparietal cortex (motor) | |
| Laminae I and II | 2+ |
| Laminae III and IV | 3+ |
| Laminae V and VI | 3− |
| Striate cortex (area 18a) | |
| Laminae I and II | 2+ |
| Laminae III and IV | 4− |
| Laminae V and VI | 3− |
| Cingulate cortex (all laminae) | 2 |
| Entorhinal cortex (all laminae) | 3− |
| Olfactory tubercle | 3 |
| Basal ganglia | |
| Caudate-putamen | 2+ |
| Globus pallidus | 1 |
| Claustrum | 4− |
| Septal areas | |
| Nucleus accumbens ) | 3 |
| Nucleus of the diagonal band (Broca) | 2 |
| Lateral septal nucleus | 1+ |
| Medial septal nucleus | 1+ |
| Triangular septal nucleus | 1 |
| Bed nucleus of the stria terminalis ) | 1+ |
| Amygdala | |
| Basolateral nucleus | 3− |
| Central nucleus | 2+ |
| Medial nucleus | 2+ |
| Thalamus | |
| Lateral thalamus | 2 |
| Medial thalamus | 1+ |
| Anterodorsal thalamic nucleus | 2+ |
| Habenula (medial plus lateral) | 2 |
| Hypothalamus | |
| Lateral hypothalamus | 1+ |
| Medial hypothalamus | 1+ |
| Anterior hypothalamus | 2− |
| Paraventricular nucleus | 2 |
| Arcuate nucleus | 0 |
| Mammillary peduncle | 3− |
| Median eminence | 4+ |

**Table 5.5 (Continued)**

| *Region* | *Relative density of CRF receptors* |
|---|---|
| Hippocampus (whole) | 1+ |
| Dentate gyrus | 1+ |
| Subiculum | 2− |
| Corpus callosum | 0 |
| Forces minor corpus callosum | 0 |
| Fornix | 0 |
| Anterior commissurea | Variable (1) |
| Optic chiasm | 0 |

## Vasopressin

The nonapeptide, arginine vasopressin VP(Arg), is the mammalian antidiuretic hormone with well-established physiological roles such as *antidiuresis* in kidney, stimulation of vascular smooth muscles to elevate blood pressure, the *pressor* effect and *glycogenolysis* in liver. The best known VP(Arg) immunoreactive brain cells are found in the *hypothalamo–neurohypophyseal system*, which releases the hormone into the bloodstream to exert hormonal effects in the periphery. However, VP(Arg) is also synthesized at multiple sites in the brain and these neurons project to different brain regions where VP(Arg) may function as a neurotransmitter and/or neuromodulator. These findings implicate VP(Arg) in processes such as of consolidation memory, temperature regulation, sexual behaviour and blood pressure regulation[49].

VP(Arg) has been studied for many years in relation to its effects on learning and memory[50]. In the initial studies VP(Arg) and its related peptides have been shown to maintain the conditioned avoidance response for long periods in rats, which has been interpreted as an enhancement of memory consolidation[51,52]. More recent studies have demonstrated VP-induced appetite motivated responses (non-aversive), effects on arousal and motivation, and an enhancement of memory retrieval in animals treated with amnestic agents[50,53].

In the past decade, several very potent VP(Arg) agonists and antagonists which can influence peripheral VP effects have been synthesized (see Table 4.5) and many of them appear to influence central processes as well[49,54,55]. Before 1982, VP(Arg) receptors at the periphery were characterized using $^{3}H$-labelled VP(Lys) for kidney, liver and smooth muscles. In the kidney a VP(Arg)-sensitive adenylate cyclase has been demonstrated, indicating that VP(Arg) second messenger is cyclic AMP, and the binding site has been termed as *$V_2$-receptor*. On the other hand, the mechanism by which VP(Arg) activates hepatic glycogenolysis and smooth muscle contraction is quite different, and involves $Ca^{2+}$ mediated processes and the participation of

**Table 5.6** Distribution of endogenous IR-SRIF and density of binding sites of [$^{125}$I]SRIF in the rat brain (modified according to Srikant and Patel(1981)[48])

| Brain region | IR-SRIF (ng/mg protein) (1) | [$^{125}$I]SRIF binding to membrane synaptosomes (fmol/mg protein) (2) | Ratio (2)/(1) |
|---|---|---|---|
| Cortex | 6.72 ± 1.70 | 155.0 ± 17.6 | 23 |
| Thalamus | 2.90 ± 0.30 | 89.5 ± 6.0 | 31 |
| Hypothalamus | 36.70 ± 2.00 | 79.0 ± 4.3 | 2 |
| Striatum | 2.34 ± 0.90 | 61.5 ± 4.8 | 26 |
| Medulla/pons | 3.27 ± 0.30 | 1.8 ± 0.15 | 0.6 |
| Cerebellum | 0.51 ± 0.08 | ~0.1 ± | 0.02 |

guanine nucleotide binding protein with the breakdown of phosphatidylinositol, and these actions are mediated via the *$V_1$-receptor type*[56,57].

$V_1$ and $V_2$ receptors have also been characterized in the anterior pituitary in brain tissue. Thus ligand binding studies indicated that the characteristics of adenohypophyseal VP(Arg) receptors are distinct from those of either $V_1$ or $V_2$ receptors[58]. This subtype is designated as $V_{1b}$ based on a marked difference in affinity for VP(Arg) antagonists such as $d(CH_2)_5$Tyr(Me)VP(Arg). $V_{1b}$-receptors mediate VP(Arg)-induced potentiation of ACTH release from adenohypophysis caused by CRF, and it has been shown that this effect is mediated via phosphatidylinositol 4,5-biphosphate metabolism in the anterior pituitary, which is consistent with the classification of these receptors as $V_1$[59].

Vasopressin receptors in the brain have been more difficult to demonstrate, most likely due to the rapid metabolism of peptide in the brain tissue. Ligand binding studies have shown VP(Arg) binding sites in the synaptic membranes of hippocampus and brain stem[60–63], and binding sites in these and other brain regions have been identified also by autoradiography[64]. An introduction of VP(Arg) analogues with low peripheral activity, such as (des-9-glycinamide) VP(Arg) (DGAVP) and (des-9-glycinamide, 8-lysine)$VP_1$(DGLVP), as well as $V_1$-specific antagonists $d(CH_2)_5$Tyr(Et) (VAVP) and $d(CH_2)_5$Tyr(Me)VP(Arg) and $V_2$-specific antagonist [$d(CH_2)_5$(2-D-Ile,4-ILe)e]VP(Arg)[65], and $V_1$-specific agonist, (2-Phe,3-Ile,8-Orn)VP and two agonists for $V_2$-receptors (1-deamino,4-Val,8-D-Arg)VP and (1-deamino,8,D-Arg)VP[50] have helped considerably to characterize vasopressin receptors in the brain. In contrast studies involving ($^3$H)VP(Arg), receptor autoradiography revealed a much more selective binding pattern with lower background labelling[66]. Brain regions such as the lateral septum, central amygdala and dorsal hippocampus were shown to be rich with so-called brain $V_{1a}$ subtype receptors, and which are also found in the lateral habenula, suprachiasmatic nucleus, arcuate nucleus and pineal gland.

## Oxytocin

Another neurohypophyseal hormone is the nonapeptide *oxytocin*, which like VP(Arg) is transported axonally from the hypothalamus to the posterior pituitary. This hormone, when released into the blood, causes peripheral effects such as uterine contraction, during parturition and contraction of myoepithelial cells in the mammary gland, and underlies the mechanism of *lactation*. Radioimmunoassay and immunocytochemical studies have revealed that oxytocin is widely distributed within the brain in the extrahypothalamic regions, including motor cortex, medulla oblongata, hippocampus, striatum/septum containing the septal nuclei, the bed nucleus of the stria terminais, the caudate nucleus, the putamen and the pallidum[67]. Work with hippocampal synaptic membranes has revealed that oxytocin has two types of binding sites[63]: the first binding site has a very low affinity to oxytocin but displays much higher activity to VP(Arg), while the second binding site, the so-called oxytocin receptor, displays equal affinity to VP(Arg) and oxytocin, and this has also been shown to be true for several other brain areas[55,63].

A newly developed VP(Arg) $V_1$-receptor antagonist $d(CH_2)_5Tyr(Me)$ VP(Arg) can discriminate between VP and oxytocin binding sites in neuronal membranes. The use of this antagonist, in receptor autoradiography studies, indicates that areas such as olfactory nucleus, ventromedial hypothalamic nucleus and the ventral hippocampus, which had been previously reported to display VP(Arg) binding sites, are either unlabelled or only labelled to a low level and these results may represent a predominant distribution of the oxytocin receptors[55]. Such a distribution suggests that oxytocin may act as a putative neurotransmitter and may be involved in different aspects of behaviour.

Although various stimuli are known to regulate the content and release of oxytocin and VP(Arg) via the hypothalamo–neurohypophyseal axis, little evidence is available with regard to the physiological events which control the levels of these peptides in the extrahypothalamic brain loci. Using the footshock brightness discrimination test in rats, a group of animals were selected with a high behavioural performance; the animals were then used in combination with a highly specific and sensitive RIA to determine the concentration of oxytocin and VP(Arg) in different brain areas. These results suggest that high endogenous levels of both neuropeptides in the septum/striatum and low level oxytocin in the hippocampus may play a key role in the performance of this conditioned reaction[67]. A functional dependence between content of both oxytocin and VP(Arg) in the central pathways and oscillation of gonadal steroids during cycling in female rat has also been shown to be independent of changes at the periphery[69].

## Delta-sleep-inducing peptide

The nonapeptide, delta-sleep-inducing peptide (DSIP), was isolated, characterized and synthesized a decade ago from the 'sleep dialysate', i.e. the blood of donor animals kept asleep by electrical stimulation of the ventromedial intralaminar thalamus[70]. DSIP was shown to induce mainly 'delta' or 'slow wave' sleep in rabbits, rats, mice and humans, whereas in cats the effect on REM sleep was more pronounced[71]. DSIP distribution within the central nervous system has been determined using both a RIA and immunocytochemistry, while the content of DSIP in body fluids and peripheral tissues has been determined by the RIA. Immunoreactive (IR) DSIP neurons were shown to be widespread in the rat brain[72], extending as a continuous rostral–caudal band in the ventral one-third of the forebrain, from the primary olfactory cortex to the lateral hypothalamus. Immunoreactive DSIP was also present in the basal ganglia, amygdala, septum, thalamus, brain stem including the reticular formation, raphe nuclei, inferior colliculus, cerebellum, locus ceruleus, periventricular gray matter and the vagal and hypoglossal nuclei.

The immunocytochemical analysis revealed that the DSIP peptide was found almost exclusively in neuronal perikarya, while immunoreactive fibres were demonstrated only near the third ventricle. This finding is somewhat different in comparison to other peptides which are predominantly located in the nerve terminals and fibres. Treatment with colchicine increased the number of neurons seen near the ventricle, which confirmed that DSIP is, like the other peptides, synthesized in the cell bodies and transported in axons. An explanation for the inability to demonstrate (IR) DSIP in nerve fibres may be in its possible association with a large protein, which makes the molecule unrecognizable to DSIP antisera[72]. Several neuropeptides are believed to exist in multiple forms, as for instance SRIF as prosomatostatin, which is a 28 to 25 amino acid form, or oxytocin and VIP in conjunction with neurophysins[30]. It is noteworthy that anti-DSIP antibodies, used for the immunocytochemical brain DSIP localization, cross-react with the DSIP metabolite des-Trp-DSIP (73%) which is an octapeptide identical to DSIP but without N-terminal tryptophan. Cross-reaction is obtained also with the synthetic analogue N-Tyr-DSIP (80%) or with nonapeptide having substituted amino acid within the molecule, as for example, D-alanine instead of glycine in third and fourth position (100%). However, less than 0.1% cross-reactivity was seen with fragments 4-6, 4-7, 4-8 and 6-9, while about 3% is obtained with pentapeptide DSIP 4-9.

The function of DSIP is not completely understood and its role in sleep is still under investigation[73]. However, the immunocytochemical distribution as well as electrophysiological and behavioural experiments suggest that DSIP may act as neurotransmitter and/or neuromodulator in several central nervous system pathways, such as visual, auditory, vestibular, as well as taking part in the arousal system and locomotor pathways.

## Insulin

The large polypeptide, pancreatic hormone insulin, secreted by the $\beta$ cells of the islets of Langerhans, plays an important role in the regulation of the uptake of glucose and other metabolites by adipose tissue, skeletal muscle and liver. In the last few years, besides this well characterized peripheral hormonal role, it has been established that the central nervous system represents a target organ for insulin. The identification of insulin and its receptors in the mammalian brain has lead to the inclusion of this peptide in the family of gastroenteropancreatic neuropeptides[74].

Conflicting results have been obtained when the insulin concentrations in brain extracts were determined by different radioimmunoassay procedures. The first studies suggested that the concentration of immunoreactive insulin (IRI) in the brain was some 25 fold higher than basal plasma insulin levels[75], which was not confirmed in later reports, when much lower values were obtained, i.e. less than 1 ng/g measured as extractable brain IRI in the rat[76,77]. These disparities have been attributed to different methods of extraction as well as to different insulin antibodies and standards used in RIA. IRI in acid-ethanol extracts of the rat brain is immunologically similar to pancreatic insulin, eluting with a molecular weight close to the pancreatic insulin, and exhibiting a similar retention time when analysed by HPLC[78,79]. The concentration of IRI varies depending on the brain region, but the olfactory bulbs and hypothalamus have the highest content[80].

Investigations of the insulin receptors in the central nervous system have provided insight into the location of the possible sites of action of this peptide and its role in brain function[74]. Insulin receptors are concentrated in the olfactory bulb of the rat brain, and relatively high binding was observed in the limbic system as well as associated with membranes isolated from hypothalamus, cerebral cortex, cerebellum and hippocampus[81–84]. Conventional binding studies have a low anatomical resolution, giving only the average concentration of peptide receptors in a block of tissue such as the olfactory bulb or hypothalamus. These findings do not represent the true distribution of receptors in the tissue, which may be specifically localized to a small region within the block. The cellular location of brain insulin receptors has also been determined by autoradiography with [$^{125}$I]insulin *in vivo*, and localized to the circumventricular organs which lack a blood–brain barrier. This finding is consistent with the hypothesis that plasma insulin action on the brain may be mediated via neuronal endings in the circumventricular organs[85].

Localization of insulin receptors in brain slices by quantitative autoradiography (QAR) gave an estimate of the number of insulin receptor binding sites in morphologically discrete foci, particularly in the limbic system and cortical layers[86]. Quantitative measurements enhanced by computer image processing and densitometry showed high concentrations of insulin binding sites in the external plexiform layer of the olfactory bulb[87]. It is tempting to propose that insulin may modulate synaptic transmission in the external plexiform layer, thus influencing the integration of olfactory stimuli, which in turn can modify food intake behaviour. In the rat hypothalamus, insulin receptors are concentrated in the arcuate, dorsomedial and paraventricular nuclei[86]. Such a distribution suggests that insulin may influence the synthesis and/or release of peptides present in these locations, which might imply that insulin exerts a modulating role on the physiological processes regulated by hypothalamic hormones. A relatively high density of insulin receptors has been found in the cerebellum, amygdala, inferior olive and nucleus of the trigeminal tract[86,88]. However, the functional significance of insulin receptors in these regions is still unknown, but the binding studies suggest that insulin may have a neuromodulatory role interacting with other neurotransmitters.

The structure and biological properties of insulin are similar to those of so-called insulin growth factors (IGFs), and the presence of IGFs receptor binding sites in the brain[89] pose the question as to whether there is an overlap between receptors for insulin and IGFs in the brain. Although membrane binding data have demonstrated that areas in the limbic system possess a high density for both insulin and IGFs receptors, quantitative autoradiographic techniques have revealed that the anatomical locations for insulin and IGFs in the limbic system are different[90]. A similar situation is found in the hypothalamus, where insulin binding sites are concentrated in the arcuate, dorsomedial and paraventricular nuclei[86], while IGF-I receptors are concentrated in the median eminence[91]. In the olfactory lobes IGF-I binding sites are concentrated in the glomerular and nerve fibre layer[91], whereas insulin receptors are located in the external plexiform layer[87]. In the cerebral cortex IGF-I receptors have the highest density in layers II and VI, while insulin receptors are concentrated in layers III and IV[74]. Such separate anatomical distribution of receptors for insulin and IGFs in the brain is consistent with the hypothesis that these peptides have different physiological roles in the brain.

## Angiotensin II

The physiological role of the renin–angiotensin system in the regulation of the arterial blood pressure and salt–volume homoeostasis is well established and mediated via its *peripheral actions*. This includes the regulation of total peripheral resistance, aldosterone release and drinking, and via synergistic *central actions* which includes central pressor effects, antidiuretic hormone release and drinking[30,92]. The biologically active peptides within the renin–angiotensin system are *angiotensins II* and *III*. Angiotensin II is an octapeptide which may be formed both in the periphery and within the central nervous system from its precursor decapeptide *angiotensin I*. This precursor is not biologically active, but represents the substrate for the enzyme *dipeptidylcarboxypeptidase* (converting enzyme) which cleaves two C-terminal amino acids from angiotensin I, manufacturing angiotensin II. Angiotensin I is formed from slightly larger tetradecapeptide, *angiotensinogen* which in fact represents the renin-substrate. Renin is the enzyme present in the circulating blood, synthesized mainly by the kidney and probably some other organs, including brain and uterus, which hydrolyses angiotensinogen to angiotensin I in the periphery. An equivalent enzyme to renin, having the same *endopeptidase* activity, is present in the brain[93]. Thus both periphery and brain have the metabolic pathways for transformation of angiotensin. *Angiotensin III*, which is biologically active, is manufactured from angiotensin II, and this may be rapidly metabolized by an *aminopeptidase* from the synaptosomal preparations. Both intravenous and intracerebroventricular administrations of angiotensin II and III cause, in different species, dipsogenic and pressor effects, natriuresis, antidiuresis, salt appetite and lowering of the plasma and CSF sodium level[94].

Angiotensin concentration has been determined in the brain tissue by chromatographic analysis and the lowest concentration was found in the cerebrum (60–70 ng/g), whilst the diencephalon has the highest content (140–140 ng/g); all other areas, except the cerebellum contained more than cerebrum[95]. Angiotensin II concentration was determined in the rat brain as immunoreactive (IR) angiotensin II using radioimmunoassay and never more than 80 pg/g was found[96]. However, the analysis of the eluting patterns from an IR angiotensin II molecular sieving column failed to demonstrates the authentic peptide, which indicates that the activity was in a higher molecular weight fraction. Although the concentration of angiotensin II is difficult to measure within the brain, the presence of angiotensinogen and all the enzymes necessary for proteolytic conversion of angiotensins, strongly suggests that angiotensin II may be released from a distinct area of the brain.

In order to determine sites within the brain at which angiotensin II may act to cause its well-known effects, mapping of the distribution of its binding receptors sites was performed by *in vitro* autoradiography in the sheep brain using [$^{125}$I](Ser$^1$,Ile$^8$) angiotensin II[97]. The highest binding density was obtained in frozen sections of the sheep brain stem in the organum vasculosum of the lamina terminalis (OVLT) and subfornical organ (SPO), while considerable binding was also found in the median preoptic nucleus (MnPO), dorsal motor nucleus of the vagus and nucleus tractus solitaris (NTS). Weaker binding was observed in the supraoptic nucleus (SON) and paraventricular nuclei (PVN) of the hypothalamus, and in the area postrema, cerebellum and posterior ventral thalamus.

The major neural connections of the MnPO and angiotensin binding sites in the brain were investigated in the sheep and rat by tracing the anterograde and retrograde transport of wheat germ agglutinin conjugated with horse-radish peroxidase (WGA-HRP) which had been microinjected into the MnPO[94] (Figure 5.1). Retrograde transport was observed into the SFO and OVLT and also into the parabrachial nucleus, NTS and ventrolateral medulla, which indicates that these regions may send the monosynaptic inputs to the MnPO. Anterograde transport of the WGA-HRP tracer was

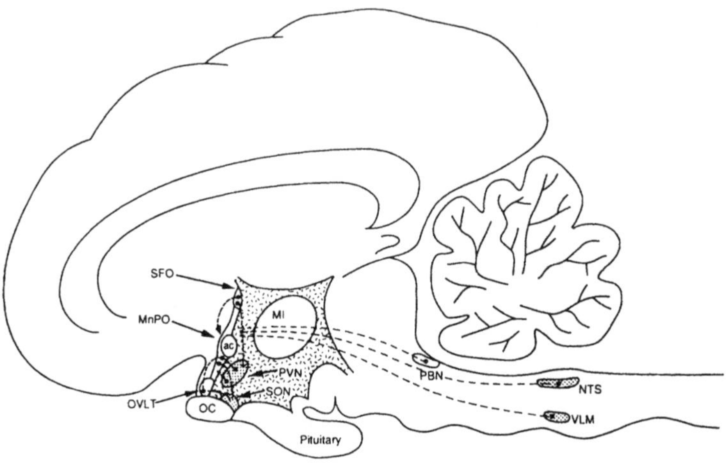


**Figure 5.1** Schematic representation of the major neural connections of the MnPO and angiotensin II binding sites in the brain studied by tracing the transport of WGA-HRP into the MnPO (according to McKinley *et al.*(1987)[94]. Mi, massa intermedia; OC, Optic chiasma; PBN, parabrachial nucleus; VLM, ventrolateral medulla; other abbreviations in the text

observed in the SON and PVN along with a number of other hypothalamic and septal sites, suggesting that neurons in the MnPO project to the SON and PVN. These findings suggest that there are monosynaptic links between many of the regions exhibiting high levels of angiotensin II receptors, which raises the possibility of the existence of a network of angiotensinergic neurons.

In order to study the physiological role of angiotensin II receptor brain zones, experimental lesions were performed with ablation of the lamina terminalis, which is the richest receptor field for this peptide in the sheep brain[98]. Such a lesion completely destroys the midline MnPO and OVLT, and causes the animal to be adipsic or hypodipsic, while the response to thirst and vasopressin release are also disrupted, and the animals exhibit impaired excretion of sodium in response to dehydration[99]. Based on these and previous results showing that i.c.v. injection of angiotensin II causes thirst, natriuresis and antidiuresis, it may be supposed that angiotensin II receptors at the lamina terminais may be a part of the neuronal system involved in the homoeostatic responses to *hypertonicity* and *dehydration*. The receptors at the circumventricular organs such as the SFO and OVLT may be sensitive to blood-borne angiotensin II, while those in the MnPO and pariventricular nucleus may respond to centrally released angiotensin II.

## Bombesin-like peptides

Bombesin is a tetradecapeptide originally isolated from amphibian skin, and subsequently homologues of this peptide were found to be located in the brain, gut and skin of mammals[100]. Although bombesin-like peptides involve a family of several peptides, such as its 27 amino acid mammalian relative, gastrin-releasing peptide (GRP), alytensin, litorin and ranatensin, the carboxyl terminal nonapeptide amino acid sequence remains identical in all these peptides, and this sequence is the major determinant of biological activity; variation in amino acid sequence was found in the N-terminal portions of these molecules. Bombesin-like peptides can be extracted from the mucosa and the muscle layers of the mammalian stomach and the muscle layers of the small intestine[101]. Here these peptides exhibit a characteristic stimulatory effect on gastrointestinal and urinary smooth muscles, as well as stimulating the release of several gastrointestinal hormones, including gastrin.

Crude dissections of rat brain tissue reveal the highest concentration of bombesin-like immunoreactivity in the hypothalamus, and the lowest in the cerebellum. The distribution of bombesin receptor binding sites generally parallels that of the extractable peptide except for hippocampus, which had a high receptor density and the lowest peptide content[30]. Most of the antibodies used for radioimmunoassay of bombesin have been directed to the C-terminal portion of the molecule, thus binding to the biologically active portion of bombesin fragments.

Direct demonstration of bombesin receptors in rat brain membranes has been achieved by the use of a radioreceptor assay with $Tyr^4$-bombesin, and a single class of high-affinity receptor binding sites ($K_d = 3$ nmol/l) was found. Mapping of the bombesin receptors revealed a high concentration in the limbic forebrain, hippocampus, amygdala, hypothalamus, periaqueductal gray matter, caudate-putamen and forebrain[102]. Apart from the direct peripheral nervous regulation of gastrointestinal motility, the stimulation of secretion and the release of hormone by bombesin, it seems that at least part of these effects are mediated also *centrally*, since they are dependent upon an intact hypothalamo–pituitary–adrenal axis[103]. On the other hand, in addition to the central effects on body temperature, on blood glucose, on sleeping and feeding behaviour, on locomotor activity and the generalized arousal effects of bombesin[104], behavioural pharmacological experiments have suggested that grooming/scratching behaviour may be mediated by nociceptive mechanisms[105]. This has been supported by the finding that the phenomenology of the behavioural actions of bombesin may involve interactions with the opiate system[106,107], which suggests that a neuromodulator role of brain bombesin-like peptides may be included in the mechanisms of *nociceptive* integration.

## Cholecystokinin

Two pancreatic hormones, hexadecapeptide, gastrin, and the 33 amino acid peptide, cholecystokinin (CCK), have for a long time been established as peripheral local hormones which play an important physiological role in the regulation of the gastrointestinal activity. Both peptides have in common the C-terminal pentapeptide, gly-trp-met-asp-phe-$NH_2$ on which the majority of the biological activity is dependent.

By means of sequence specific radioimmunoassays it has been demonstrated that CCK is widely distributed in the central nervous system. The highest concentration is found in the neocortex (0.2–2.3 nmol/g), in the striatum (0.2–0.5 nmol/g) and in the hypothalamus (up to 0.2 nmol/g), while there was no CCK in the pineal gland, pituitary, cerebellum, pons and spinal cord[9,108]. Only 5% of the total immunoactive material consisted of $CCK_{33}$, with more than 80% being C-terminal octapeptide $CCK_8$.

A number of studies have been performed to characterize high affinity binding sites for CCK in the brain, both by ligand binding to membrane particles, and by *in vitro* quantitative autoradiography. It has been shown in the rat and mouse that the regions with highest specific binding for CCK are found in the cerebral cortex, olfactory bulb, caudate nucleus, hypothalamus and hippocampus[109–111]. A comparative study was made of CCK receptors in the brain of four different species, namely mouse, rat, hamster and guinea-pig, using $[^{125}I]CCK_{33}$ as a ligand and conventional binding procedure to the brain membranes. In these four species the same order of potency was found for the competition between unlabelled analogues and the radiolabelled ligand for the binding sites, namely $CCK_8 > CCK_{33} >$ desulphated $CCK_8 > CCK_4$[112]. According to behavioural experiments and receptor CCK localization in the brain it has been postulated that this CCK system within the brain may modulate *feeding behaviour* and is intimately connected with *satiety*.

## Vasoactive intestinal peptide

The octacosapeptide, vasoactive intestinal peptide (VIP), initially isolated from the gastrointestinal tract[113] was shown to cause systemic vasodilation with a consequent hypotensive effect, an increased cardiac output and hyperglycaemia, many of these actions being associated with stimulation of adenylate cyclase and the cAMP[104].

VIP is also widely distributed within the central nervous system, the highest concentration, in man and pig, being found in the hypothalamus, hippocampus and frontal cortex, while substantia nigra, cerebellum and thalamus exhibit a much lower concentration of VIP[114]. VIP receptors have been characterized using membrane binding studies and autoradiographic visualization *in vitro*. The discrete regional distribution of $[^{125}I]$VIP binding in rat brain slices by autoradiography has revealed that VIP receptors are confined to certain areas of gray matter, but not the white matter[115]. Particular high densities are found in the dentate gyrus, pineal gland, supraoptic and suprachiasmatic nuclei, superficial gray layer of the superior colliculus and in the area postrema. Moderate grain densities were also present in the olfactory bulb and tubercle, cerebral cortex, nucleus accumbens, caudate-putamen, interstitial nucleus of the stria terminalis, paraventricular thalamic nucleus, medial amygdaloid nucleus, subiculum and the medial geniculate nucleus, while grains were absent in the corpus callosum[115].

These moderate to high concentrations of VIP receptors in certain brain regions are an indication that VIP may play a role in integrating sensory information in the *visual* and *olfactory* systems, and in regulating the brain *microcirculation*.

## BIOSYTHESIS OF BRAIN PEPTIDES

Various peptides within the brain have been identified, localized, chemically characterized, and in the most recent studies considerable care has been taken to confirm the authenticity of the material described. However, one of the key questions concerned with the biosynthesis of these peptides, is whether the peptide is synthesized *de novo* in the brain structure from which it was found, or did the peptide originate from outside the CNS? A direct answer to this question may be obtained by the demonstration of synthesis of the peptide in the brain and this can be achieved by employing one of the following approaches[1,116]. Firstly, it is necessary to demonstrate accumulation of a given peptide in cells and culture media during *in vitro* isolation of a certain brain region. Secondly, one can demonstrate incorporation of labelled amino acid into the peptide secreted from the brain tissue. Thirdly, one may use a cDNA probe prepared from mRNA obtained from non-neural tissue with an aim of identifying the mRNA of the neural tissue, and vice versa, by extracting mRNA from neural tissue following by translation into a cell-free system one may characterize peptide products of this translation. Also by means of *in situ* hydridization experiments with labelled cDNA, the cellular site of synthesis for a given peptide may be localized[1,117]. For some neuropeptides a significant body of evidence is available in favour of *de novo* biosynthesis *in situ* within the brain, while for the others the specific secreting neural cells have yet to be positively identified.

It has been well established that most biologically active peptides are synthesized in the brain from large *precursors* involving a cascade of enzymatic processes which determine the quantity and potency of synthesized neuropeptides. An initial step represents a transcription of a gene into the *peptide precursor* RNA, which by splicing may result in formation of different mature mRNAs from a single gene. The consequence of this step results in a set of different neuropeptides. For instance, calcitonin (CT) and CGRP are derived from the same gene, but in the thyroid tissue the mRNA encodes CT, while in the brain different mRNA encodes CGRP[118]. Similarly, one gene encodes different mRNAs responsible for the production of substance P and various tachykinins[119]. On the other hand, processing of neuropeptides from the same precursor can also be cell-specific as with pro-opiomelanocortin (POMC)-peptides. POMC is predominantly processed to ACTH and $\beta$-LPH in the anterior lobe of the hypophysis, and to $\alpha$-MSH and $\beta$-E in the intermediate lobe of the hypophysis[1,116]. Enzymatic processing of this molecule in the brain tissue resembles that in the intermediate lobe except that a much smaller percentage of the peptide products are acetylated. POMC peptides may undergo various non-proteolytic modifications such as acetylation, sulphation, glycosylation and phosphorylation which influences their biological activity.

The extent of brain peptide processing depends to a large extent on enzymatic cleavage from the precursor at 'signalling' pairs of basic amino acids, followed by a *carboxypeptidase B* type of enzymatic cleavage of the basic amino acid attached to the carboxyl terminal. Probably enkephalin- and angiotensin-synthetic systems in the brain are the best studied examples which support this view. Thus, for instance, in the *enkephalin processing system*, one of the early precursor fragments isolated from the brain tissue is $\alpha$-neo-endorphin which consists of 15 amino acids with a group of three basic amino acids at the carboxyl terminal group of Enk(Leu)[15]. In addition, a carboxypeptidase type of enzyme has been isolated from brain tissue and termed '*enkephalin convertase*', which is supposed to be selective for the physiological enkephalin biosynthesis to convert enkephalin precursors to enkephalins. In the *angiotensin processing system* a subcellular localization of the proteolytic enzymes, *endopeptidase*, *dipeptidylcarboxypeptidase* and *amino peptidase* in the brain has been demonstrated[93] and the precursor fragments were also isolated, confirming the existence of a similar type of biosynthesis as in the enkephalin system.

On the other hand, for some peptides, there is still controversy about their *de novo* brain synthesis. For example in the case of *insulin*, the support for the hypothesis of neural insulin synthesis includes the results of immunocytochemical studies, both *in situ* and in culture, but the cross-reactivity of insulin antibodies with insulin-like growth factors makes a clear interpretation difficult[74]. Also reports of immunostaining of IRI in cultured neurons must be viewed with some caution since neurons may produce multiple forms of IRI which have been shown by HPLC to be different from pancreatic insulin.

It has been shown that most of the brain peptides can also be synthesized outside the brain, in the pituitary, gastrointestinal tract, lung and other tissues, from where they can enter the general circulation, representing the *extra-brain* pool of *circulating peptides*[120]. Some peptides are also released in the CSF from hypothalamic loci; thus they may obtain access to the central nervous system via specific ependymal tanacyte transport, or via the permeable ependyma. Although most of the circulating peptides have relatively short half-lives in the blood, there is still a possibility that blood-borne peptides may reach the target areas in the central nervous system from this peripheral pool. These peptides have to interact with the brain capillary endothelial wall, which forms the basis of the blood–brain barrier. Blood-borne peptides may also enter the brain and CSF at points where the blood–brain barrier is absent at the level of the cerebral capillaries, and influence other regions of the brain by interaction with receptors within these regions, namely at the circumventricular organs. There is a barrier to the free diffusion of large molecules from these areas into the CSF located at the level of the ependyma in these organs. On the other hand, *brain-derived peptides*, which are released from peptidergic neurons during specific brain activity, pass directly into the brain extracellular fluid and/or CSF. These peptides may also interact with the blood–brain, blood–CSF barrier and circumventricular organs, which may influence their metabolic turnover and fate in the central nervous system.

## References


1. Krieger, D.T. (1986). An overview of neuropeptides. In: *Neuropeptides in Neurologic and Psychiatric Diseases*, eds., J.B. Martin and J.D. Barchas, Raven Press, New York, 1–32
2. Mutt, V. (1982). Gastrointestinal hormones: a field of increasing complexity. *Scand. J. Gastroenterol. (suppl)*, **77**, 133–152
3. Bayliss, W.M. and Starling, E.H. (1902). The mechanism of pancreatic secretion. *J. Physiol.*, **28**, 325–353
4. Pearse, A.G.E. (1976). Peptides in brain and intestine. *Nature (London)*, **262**, 92–94
5. Pearse, A.G.E. (1979). The APUD concept and its relationship to the neuropeptides. In: *Brain Peptides: A New Endocrinology*, eds., A.M. Gotto, Jr., E.J. Peck, Jr. and A.E. Boyd, III. Elsevier North-Holland, Amsterdam, 89–101
6. Vanderhaeghen, J.J., Signeau, J.C. and Gepts, W. (1975). New peptide in the vertebrate CNS reacting with antigastrin antibodies. *Nature (London)*, **257**, 604–605
7. Krieger, D.T. and Martin, J.B. (1981). Brain peptides. *N. Engl. J. Med.*, **304**, 876–885
8. Krieger, D.T. (1983). Brain peptides: what, where and why? *Science*, **222**, 975–985
9. Figlewicz, D.P., Lacour, F., Sipols, A., Porte, Jr., D. and Woods, S.C. (1987). Gastroenteropancreatic (GEP) peptides in the central nervous system. *Ann. Rev. Physiol.*, **49**, 383–395
10. Heidenreich, K.A., Zahniser, N.R., Berhanu, P., Brandenburg, D. and Olefsky, J.M. (1983). Structural differences between insulin receptors in the brain and peripheral target tissues. *J. Biol. Chem.*, **258**, 8527–8530
11. Kastin, A.J., Olson, R.D., Fritschka, E. and Coy, D.H. (1981). Neuropeptides and the blood–brain barrier. In: *Cerebral Microcirculation and Metabolism*, eds., J. Cervos-Navarro and J. Fritschka. Raven Press, New York, 139–145
12. Kastin, A.J., Schally, A.V., Olson, G.A. and Coy, D.H. (1983). Early questions involving hypothalamic peptides. In: *The Anterior Pituitary Gland*, ed., A.S. Bhatnagar. Raven Press, New York, 55–59
13. Hughes, J., Smith, T.W., Kosterlitz, H.W., Fothergill, L.A., Morgan, B.A. and Morris, H.R. (1975). Identification of two related pentapeptides from the brain with potent opiate agonist activity. *Nature (London)*, **258**, 577–579

14. Simantov, R. and Snyder, S.H. (1976). Morphine-like peptides in mammalian brain: isolation, structure, elucidation, and interactions with opiate receptors. *Proc. Natl. Acad. Sci. USA*, **73** (7), 2515–2519
15. Goodman, R.F., Fricker, L.D. and Snyder, S.H. (1983). Enkephalins. In: *Brain Peptides*, eds., D.T. Krieger, J. Browstein and J.B. Martin. J Liley & Sons, New York, 828–849
16. Pickel, V.M., Sumal, K.K., Beckley, S.C., Miller, R.J. and Reis, D.J. (1980). Immunocytochemical localization of enkephalin in the neostriatum of rat brain: a light and electron microscopic study. *J. Comp. Neurol.*, **189**, 727–940
17. Paterson, S.J., Robson, L.E. and Kosterlitz, H.W. (1983). Classification of opioid receptors. *Br. Med. Bull.*, **39**, 31–36
18. Demoliou-Mason, C.D. and Barnard, E.A. (1986). Distinct subtypes of the opioid receptor with allosteric interactions in brain membranes. *J. Neurochem.*, **46**, 1118–1128
19. Demoliou-Mason, C.D. and Barnard, E.A. (1986). Characterization of opioid receptor subtypes in solution. *J. Neurochem.*, **46**, 1129–1136
20. Itzhak, Y. and Pasternak, G.W. (1986). Interaction of [D-Ser$^2$, Leu$^5$] enkephalin-Thr$^6$ (DSLET), a relatively selective delta ligand, with $\mu_1$ opioid binding sites. *Life Sci.*,**40**, 307–311
21. Oka, K., Kantrowitz, J.D. and Spector, S. (1985). Isolation of morphine from toad skin. *Proc. Natl. Acad. Sci. USA*, **82**, 1852–1854
22. Goldstein, A., Barrett, R.W., James, I.F., Lowney, L.I., Weitz, C.J., Knipmeyer, L.L. and Rapoport, H. (1985). Morphine and other opiates from beef brain and adrenal. *Proc. Natl. Acad. Sci. USA*, **72**, 5203–5207
23. Donnerer, J., Oka, K., Brossi, A., Rice, K.C. and Spector, S. (1986). Presence and formation of codeine and morphine in the rat. *Proc. Natl. Acad. Sci. USA*, **83**, 4566–4567
24. Cardinale, G.J., Donnerer, J, Finck, A.D., Kantrowitz, J.D., Oka, K. and Spector, S. (1986). Morphine and codeine are endogenous components of human cerebrospinal fluid. *Life Sci.*, **40**, 301–306
25. Koch, Y. and Okon, E. (1979). Localisation of releasing hormones in the human brain. *Int. Rev. Exp. Path.*, **19**, 45–62
26. Jackson, I.M. and Lechan, R.M. (1983). Thyrotropin releasing hormone. In: *Brain Peptides*, eds., Krieger, D.T., Browstein, J. and Martin, J.B. John Liley and Sons, New York
27. Holtman, J.R., Buller, A.L., Hamosh, P. and Gillis, R. (1986). Central respiratory stimulation produced by thyrotropin-releasing hormone in the cat. *Peptides*, **7**, 207–212
28. Jackson, M.D. (1980). Significance and function of neuropeptides in cerebrospinal fluid. In: *Neurobiology of CSF Vol. I*, ed., J.H. Wood. Plenum, New York, 625–651
29. Silverman, A.J., Knigge, K.M., Ribas, J.L. and Sheridan, M. (1973). Transport capacity of the median eminence. III. Amino acid and thyroxine transport of the organ cultured median eminence. *Neuroendocrinology*, **11**, 107–118
30. Davson, H., Welch, K. and Segal, M.B. (1987). *The Physiology and Pathophysiology of the Cerebrospinal Fluid.* Churchill Livingstone, London
31. Weindl, A. and Sofroniew, M.V. (1981). Relation of neuropeptides to mammalian circumventricular organs. *Adv. Biochem. Pharmacol.*, **28**, 302–320
32. Wenger, T. (1987). The role of organum vasculosum of the lamina terminals in the regulation of pituitary gonadotrophic hormone secretion. *Wiss. Z. Karl-Marx Univ. Leipzig Math-Naturwiss R.*, **36** (1), 52–55
33. Schivers, B.D., Harlan, R.E., Romano, J.G., Howills, R.D. and Phaff, G.W. (1986). Cellular localization of pre-enkephalin in RNA in rat brain: gene expression in the caudate putamen and cerebral cortex. *Proc. Natl. Acad. Sci. USA*, **83**, 6221–6225
34. Vale, W., Spiess, J., Rivier, C. and Rivier, J. (1981). Characterization of a 41-residue ovine hypothalamic peptide that stimulates secretion of corticotropin and beta-endorphin. *Science*, **213**, 1394–1397

35. Rivier, C., Rivier, J. and Vale, W.W. (1982). Inhibition of adrenocorticotrophin hormone secretion in the rat by immunoneutralization of corticotropin-releasing factor. *Science*, **218**, 377–379
36. de Sousa, E.B. and Kuhar, H.J. (1986). Corticotrophin-releasing factor receptors: autoradiographic identification. In: *Neuropeptides in Neurologic and Psychiatric Disease*, eds., J.B. Martin and J.D. Barchas. Raven Press, New York, 179–198
37. Bloom, F.E., Battenburg, E.L.F., Rivier, J. and Vale, W. (1982). Corticotropin-releasing factor (CRF): immunoreactive neurones and fibers in rat hypothalamus. *Regul. Peptides*, **4**, 43–48
38. Bugnon, C., Fellmann, D., Gouget, A. and Cardot, J. (1982). Corticoliberin in rat brain: immunocytochemical identification and localization of a novel neuroglandular system. *Neurosci. Lett.*, **30**, 25–30
39. Cummings, S., Elde, R., Ells, J. and Lindall, A. (1983). Corticotropin-releasing factor immunoreactivity is widely distributed within the central nervous system of the rat: an immunohistochemical study. *J. Neurosci.*, **3**, 1355–1368
40. Swanson, L.W., Sawchenko, P.E., Rivier, J. and Vale, W.W. (1983). Organization of ovine corticotropin-releasing factor immunoreactive cells and fibers in the rat brain: an immunohistochemical study. *Neuroendocrinology*, **36**, 165–186
41. Peterfreund, R.A. and Vale, W.W. (1983). Ovine corticotropin-releasing factor stimulates somatostatin secretion from cultured brain cells. *Endocrinology*, **112**, 1275–12778
42. Sutton, R.E., Koob, G.F., LeMoal, M., Rivier, J. and Vale, W.W. (1982). Corticotropin-releasing factor produces behavioral activation in rats. *Nature (London)*, **297**, 331–333
43. Tache, Y., Goto, Y., Gunlon, M.W., Vale, W.W., Rivier, J. and Brown, M. (1983). Inhibition of gastric acid secretion in rats by intracerebral injection of corticotropin-releasing factor. *Science*, **222**, 935–937
44. Eberly, L.B., Dudley, C.A. and Moss, R.L. (1983). Iontophoretic mapping of corticotropin-releasing factor (CRF) sensitive neurones in the rat forebrain. *Peptides*, **4**, 837–841
45. Gold, P.W., Chrousos, G., Kellner, C., Post, R., Roy, A., Augerinos, P., Schulte, H., Oldfield, E. and Loriaux, D.L. (1984). Psychiatric implications of basic and clinical studies with corticotropin-releasing factor. *Am. J. Psychiatry*, **J41**, 619–623
46. Reichlin, S. (1983). Somatostatin. *N. Engl. J. Med.*, **309**, 1495–1501
47. Reichlin, S. (1983). Somatostatin. *N. Engl. J. Med.*, **309**, 1556–1562
48. Srikant, C.B. and Patel, Y.C. (1981). Somatostatin receptors: identification and characterization in rat brain membranes. *Proc. Natl. Acad. Sci. USA*, **78**, 3930–3934
49. de Wied, D., Gaffori, O., van Ree, J.M. and de Jong, W. (1984). Central target for the behavioral effects of vasopressin peptides. *Nature (London)*, **308**, 276–278
50. Hoffman, P.L., Szabo, G. and Tabakoff, B. (1988). The central effects of vasopressin and related peptides on tolerance to ethanol. In: *Peptides and Amino Acid Transport Mechanisms in the Central Nervous System*, eds., Lj Rakic, D.J. Begley, H. Davson and B.V. Zlokovic. Macmillan, 147–156
51. de Wied, D. (1973). Long-term effect of vasopressin on the maintenance of a conditioned avoidance response. *Nature (London)*, **232**, 137–143
52. van Wimersma Greidanus, Tj.B., van Ree, J.M. and de Wied, D. (1983). Vasopressin and memory. *Pharmacol. Ther.*, **20**, 437–458
53. LeMoal, M., Dantzer, R., Mormede, P., Bandel, A., LeBrun, C., Ettenbberg, A., van der Kooy, D., Wenger, J., Deyo, S., Koob, G.F. and Bloom, F.E. (1984). Behavioral effects of peripheral administration of arginine vasopressin: a review of our search for a mode of action and a hypothesis. *Psychoneuroendocrinology*, **9**, 319–341
54. Manning, M. and Sawyer, W.H. (1984). Design and uses of selective agonistic and antagonistic analogues of the neuropeptides oxytocin and vasopressin. *TINS*, **7**, 6–9
55. van Leeuwen, F.W. (1987). A new $^{3}$H-arginine vasopressin antagonist: a specific and sensitive means to show vasopressin binding in the brain, NET 945. *Biotechnology update*, **2**, 1–2

56. Abdel-Latif, A.A. (1986). Calcium mobilizing receptors, polyphosphoinositides and the generation of second messengers. *Pharmacol. Rev.*, **38**, 227–272
57. Fitzgerald, T.J., Uhing, R.J. and Exton, J.A. (1986). Solubilization of the vasopressin receptor from rat liver plasma membranes. *J. Biol. Chem.*, **262**, 16871–16877
58. Jard, S., Gaillard, R.C., Guilloni, G., Marie, J., Schoenenberg, P., Muller, A.F., Manning, M. and Sawyer, W.H. (1986). Vasopressin antagonists allow demonstration of a novel type of vasopressin receptor in the rat adenohypophysis. *Mol. Pharmacol.*, **30**, 171–177
59. Todd, K. and Lightman, S.L. (1987). Vasopressin activation of phosphatidylinositol metabolism in rat anterior pituitary in vitro and its modification by changes in the hypothalamo–pituitary–adrenal axis. *Neuroendocrinology*, **45**, 212–218
60. Dorsa, D.M., Majumdar, L.A., Petracca, F.M., Baskin, D.G. and Cornett, L.E. (1983). Characterization and localization of [$^3$H]arginine$^8$-vasopressin binding to rat kidney and brain tissue. *Peptides*, **4**, 699–706
61. Barberis, C. (1983). [$^3$H]Vasopressin binding to rat hippocampal synaptic plasma membrane. Kinetic and pharmacological characterization. *FEBS Lett.*, **162**, 400–405
62. Constantini, M.G. and Pearlmutter, A.F. (1984). Properties of the specific binding site for arginine vasopressin in rat hippocampal synaptic membranes. *J. Biol. Chem.*, **259**, 11739–11745
63. Audigier, S. and Barberis, C. (1985). Pharmacological characterization of two specific binding sites for neurohypophyseal hormones in hippocampal synaptic plasma membranes of the rat. *EMBO J.*, **4**, 1407–1412
64. Biegon, A., Terlou, M., Voorhuis, Th.D. and de Kloet, E.R. (1984). Arginine vasopressin binding sites in rat brain: a quantitative autoradiographic study. *Neurosci. Lett.*, **44**, 229–234
65. Sawyer, W.H. and Manning, M. (1985). Development of selective agonists and antagonists of vasopressin and oxytocin. In: *Vasopressin*, ed., R.W. Schrier. Raven Press, New York, 131–144
66. Caffé, A.R., Van Leewen, F.W. and Luiten, P.G. (1987). Vasopressin cells in the medial amygdala of the rat project to the lateral septum in ventral hippocampus. *J. Comp. Neurol.*, **261**, 237–252
67. Ermisch, A., Landgraf, R. and Mobius, P. (1986). Vasopressin and oxytocin in brain areas of rats with high or low behavioral performance. *Brain Res.*, **379**, 24–29
68. de Kloet, E.R., Voorhuis, D.A.M., Bosch, M.A.Y. and Elands, J. (1986). Estradiol modulates density of putative ‘oxytocin receptors’ in discrete brain regions. *Neuroendocrinology*, **44**, 415–421
69. Landgraf, R. (1987). Vasopressin and oxytocin in the rat brain: sexual differences and vesicular localisation. *Wiss. Z. Karl-Marx-Univ. Leipzig, Math.-Naturwiss. R.*, **36**, 175–178
70. Schoenenberger, G.A. (1984). Characterization, properties and multivariate functions of delta-sleep-inducing peptide (DSIP). *Eur. Neurol.*, **23**, 321–345
71. Graf, M.V. and Kastin, A.J. (1984). Delta-sleep-inducing peptide (DSIP): a review. *Neurosci. Biobehav. Rev.*, **8**, 83–93
72. Feldman, S.C. and Kastin, A.J. (1984). Localization of neurones containing immunoreactive delta sleep-inducing peptide in the rat brain: an immunocytochemical study. *Neuroscience*, **11**, 303–317
73. Susic, V. (1988). Delta sleep promoting effects of a delta sleep inducing peptide. In: *Peptide and Amino Acid Transport Mechanisms in the Central Nervous System*, eds., Lj. Rakic, D.J. Begley, H. Davson and B.V. Zlokovic. Macmillan, 141–140
74. Baskin, D.G., Dorsa, D.M., Figlewicz, D.P., Corp, E.S., Wilcox, B.J., Wallum, B.J. and Woods, S.C. (1988). Insulin as a regulatory peptide in the CNS. In: *Peptides and Amino Acids in the Central Nervous System*, eds., Lj. Rakic, D.J. Begley, H. Davson and B.V. Zlokovic. Macmillan, London. 69–81
75. Havrankova, J., Roth, J. and Brownstein, M. (1978). Insulin receptors are widely distributed in the central nervous system of the rat. *Nature (London)*, **272**, 829–831

76. Yalow, R.S. and Eng, J. (1983). Insulin in the central nervous system. In: *Adv. Metabolic Disorders, Vol. 10*, ed., A. Szabo. Academic Press, New York, 34–54
77. Baskin, D.G., Porte, Jr., D., Guest, Jr., K. and Dorsa, D.M. (1983). Regional concentrations of insulin in the rat brain. *Endocrinology*, **112**, 898–903
78. LeRoith, D., Hendricks, S.A., Lesnaik, M.A., Rishi, S., Becker, K.L., Havrankova, J., Rosenzqeig, J., Brownstein, M. and Roth, J. (1983). Insulin in brain and other extrapancreatic tissues of vertebrates and non-vertebrates. In: *Adv. Metabolic Disorders, Vol. 10*, ed., A. Szabo. Academic Press, New York, 304–340
79. Birch, N.P., Christine, D.L. and Renwick, A.G.C. (1984). Immunoreactive insulin from mouse brain cells in culture and whole rat brain. *Biochem. J.*, **218**, 19–27
80. Baskin, D.G., Stein, L.J., Ikeda, H., Woods, S.C., Figlewicz, D.P., Porte, Jr., D., Greenwood, M. and Dorsa, D.M. (1985). Genetically obese Zucker rats have abnormally low brain insulin content. *Life Sci.*, **36**, 627–633
81. Figlewicz, D.P., Dorsa, D.M., Stein, L.J., Baskin, D.G., Paquette, T., Greenwood, M., Woods, S.C. and Porte, Jr., D. (1985). Brain and liver insulin binding is decreased in Zucker rats carrying the fa' gene. *Endocrinology*, **117**, 1537–1543
82. Kappy, M., Sellinger, S. and Raizada, M. (1984). Insulin binding in four regions of the developing rat brain. *J. Neurochem.*, **42**, 198–203
83. Zahniser, N.R., Goens, M.B., Hanaway, P.J. and Vinych, J.V. (1984). Characterization and regulation of insulin receptors in rat brain. *J. Neurochem.*, **42**, 1345–1362
84. Schluter, K.J., Andre, J., Enzmann, F. and Kerp, L. (1984). Insulin receptor binding in pig brain: different affinities of porcine and human insulin. In: *Lessons from Animal Diabetes*, eds., E. Shafir and A.E. Renold. Libbey, London, 253–256
85. van Houten, M., Nance, D.M., Gauthier, S. and Posner, B.I. (1983). Origin of insulin receptive nerve terminals in rat median eminence. *Endocrinology*, **113**, 1393–1399
86. Corp, E.S., Woods, S.C., Porte, Jr., D., Dorsa, D.M., Figlewicz, D.P. and Baskin, D.G. (1986). Localization of I-insulin binding sites in the rat hypothalamus by quantitative autoradiography. *Neurosci. Lett.*, **70**, 17–22
87. Baskin, D.G., Davidson, D., Corp, E., Lewellen, T. and Graham, M. (1986). An inexpensive microcomputer digital imaging system for densitometry: quantitative autoradiography of insulin receptors with $^{125}$I and LKB Ultrofilm. *J. Neurosci. Meth.*, **16**, 119–129
88. Hill, J., Lesniak, M., Pert, C. and Roth, J. (1986). Autoradiographic localization of insulin receptors in rat brain: prominence in olfactory and limbic areas. *Neuroscience*, **17**, 1127–1138
89. Gammeltoft, S., Hasselbacher, G., Humbel, R., Fehlmann, M. and Van Obberghen, E. (1985). Two types of receptor for insulin-like growth factors in mammalian brain. *EMBO J.*, **4**, 3407–3412
90. Baskin, D.G., Figlewicz, D.P., Woods, S.C., Porte, Jr., D. and Dorsa, D.M. (1987). Insulin in the brain. *Ann. Rev. Physiol.*, **49**, 335–347
91. Bohannon, N.J., Corp, E., Wilcox, B.J., Porte, Jr., D., Figlewicz, D.P. and Baskin, D.G. (1986). Identification of specific binding sites for insulin-like growth factor-I (IGF-I) in the rat brain by *in vivo* quantitative autoradiography. *Appetite*, **7**, 243–244
92. Phillip, M.I. (1983). New evidence for brain angiotensin and for its role in hypertension. *Fed. Proc.*, **42**, 2667–2672
93. Tonnaer, J.A.D.M., Engels, C.M.H., Wiegaut, V.M., Burbach, J.P.H., De Jong, W. and De Wied, D. (1983). Proteolytic conversion of angiotensins in rat brain tissue. *Eur. J. Biochem.*, **131**, 415–421
94. McKinley, M.J., Allen, A., Clevers, J., Denton, D., Mendelsohn, A.O., Oldfield, B.J., Tarjan, E. and Weisinger, R.S. (1987). Angiotensin II receptors in the brain of sheep. *Wiss. Z. Karl-Marx-Univ. Leipzig Math-Naturwiss. R.*, **36** (2), 189–191
95. Sernia, C. and Mowchanuk, M.D. (1983). Brain angiotensin: *in vitro* synthesis and chromatographic characterization. *Brain Res.*, **259**, 275–283

96. Meyer, D.K., Phillips, M.I. and Eiden, L. (1983). Studies on the presence of angiotensin II in rat brain. *J. Neurochem.*, **38**, 816–820
97. McKinley, M.J., Allen, A., Clevers, J., Denton, D.A. and Mendelsohn, F.A.O. (1986). Autoradiographic localization of angiotensin in the sheep brain. *Brain Res.*, **375**, 373–376
98. Congiu, M., Denton, D.A., Park, R.G., Penschow, J., Simpson, J.B., Tarjan, E., Weisinger, R.S. and Wright, R.D. (1984). The anterior wall of the third cerebral ventricle and homeostatic responses to dehydration. *J. Physiol. (Paris)*, **79**, 421–427
99. Coghlan, J.P., Congiu, M., Denton, D.A., Fei, D.T. and Park, R.G. (1986). Augmented plasma renin levels in dehydrated sheep with periventricular lesions. *Brain Res.*, **376**, 416–419
100. Walsh, J.H. (1983). Bombesin-like peptides. In: *Brain Peptides*, eds., D.T. Krieger, J. Browstein and J.B. Martin. J Liley & Sons, New York, 942–960
101. Walsh, J.H., Reeve, Jr. J.R., and Vigna, S.R. (1981). In: *Gut Hormones*, 2nd edition, eds., S.R. Bloom and J.M. Polak. Churchill Livingstone, Edinburgh, 413–418
102. Moody, T.W., O'Donohue, T.L. and Jacobowitz, D.M. (1981). Biochemical localization and characterization of bombesin-like peptides in discrete regions of rat brain. *Peptides*, **2**, 75–79
103. Koslo, R.J., Gmerek, D.E. and Porreca, F. (1986). Intrathecal bombesin-induced inhibition of gastrointestinal transit: requirement for an intact pituitary-adrenal axis. *Reg. Peptides*, **14**, 237–242
104. Figlewicz, D.P., Lacourt, F., Sipol, A. and Woods, S.C. (1987). Gastroenteropancreatic peptides and the central nervous system. *Ann. Rev. Physiol.*, **49**, 393–395
105. Meisenberg, G. and Simmons, W.H. (1986). Behavioral alterations induced by substance P, bombesin and related peptides in mice. *Peptides*, **7**, 557–561
106. Van Wimersma Greidanus, Tj.B., Donker, D.K., Walhof, R., Van Grafhorst, J.C.A., DeVries, N., Van Schaik, S.J., Maigret, B.M., Spruijt, B.M. and Colbern, D.L. (1985). The effects of neurotensin haloxone and haloperidol on elements of excessive grooming behavior induced by bombesin. *Peptides*, **6**, 1179–1183
107. Van Wimersma Greidanus, Tj.B., Donker, D.K., Van Zinnicq Bergmann, F.F.M., Bekenkamp, R., Maigret, C. and Spruijt, B. (1986). Comparison between excessive grooming induced by bombesin or by ACTH: the differential elements of grooming and development of tolerance. *Peptides*, **6**, 369–372
108. Vandehaegen, J.J., Lostra, F., de Mey, J. and Gilles, C. (1980). Immunocytochemical localization of cholecystokin and gastrin like peptides in the brain and hypothalamus of the rat. *Proc. Natl. Acad. Sci. USA*, **77**, 1190–1194
109. Gaudreau, P., Quirion, R., St. Pierre, S. and Pert, C.B. (1983). Characterization and visualization of cholecystokinin receptors in rat brain using $^3$H-pentagastrin. *Peptides*, **4**, 755–762
110. Saito, A., Williams, J.A., Waxler, S.H. and Goldfine, I.D. (1982). Alterations of brain cholecystokinin receptors in mice made obese with gold thioglucose. *J. Neurochem.*, **39**, 525–528
111. Van Dijk, A., Richards, J.G., Trzeciak, A., Gillessen, D. and Mohler, H. (1984). Cholecystokinin receptors: biochemical demonstration and autoradiographical localization in rat brain and pancreas using $^3$H-cholecystokinin as radioligand. *J. Neurosci.*, **4**, 1021–1033
112. Williams, J.A., Gryson, K.A. and McChesney, D.J. (1986). Brain CCK receptors: species differences in regional distribution and selectivity. *Peptides*, **7**, 293–296
113. Said, S.I. and Mutt, V. (1970). Peptides in the broad biological activity isolated from the small intestine. *Science*, **169**, 1217–1218
114. Fahzenkrug, J. (1979). Vasoactive intestinal polypeptide: measurement distribution and putative neurotransmitter function. *Digestion*, **19**, 149–169
115. Schaffer, M.M. and Moody, T.W. (1986). Autoradiographic visualization of CNS receptors for vasoactive intestinal peptide. *Peptides*, **7**, 283–286

116. de Wied, D. (1987). The neuropeptide concept. In: *Neuropeptides and Brain Functions*, eds., E.R. de Kloet, N.M. Wiegant and D. de Wied, *Prog. Brain Res.*, Vol.72, 93–108
117. Bloom, F.E. (1987). Molecular diversity and cellular functions of neuropeptides. In: *Neuropeptides and brain function*, eds., E.R. de Kloet, W.M. Wiegnant and D. de Wied. *Prog. Brain Res.*, Vol.72, 213–223
118. Amara, S.G., Jonas, V., Rosenfield, M.G., Ong, E.S. and Evans, R.M. (1982). Alternative RNA processing in calcitonin gene expression generates mRNAs encoding different polypeptides products. *Nature (London)*, **298**, 240–244
119. Nakamuza, K., Kanako, H., Hayashi, M., Shimizu, T., Ihaza, T., Kanto, S., Kusakabe, M., Itoh, J., Yanaihaza, C., Yanaihaza, N. and Izie, M. (1986). Vasoactive intestinal polypeptide (VIP) and human cerebrospinal fluid of dementia with cerebral atrophy. *Abstracts 1st Int. Congress of Neuroendocrinology*, San Francisco, July 9–11, 1986, No. 331
120. Zlokovic, B.V., Begley, D.J., Segal, M.B., Davson, H., Rakic, Lj., Lipovac, M.N., Mitrovic, D.M. and Jankov, R.M. (1988). Neuropeptide transport mechanisms in the central nervous system. In: *Peptide and Amino Acid Transport Mechanisms in the Central Nervous System*, eds., Lj. Rakic, D.J. Begley, H. Davson and B.V. Zlokovic. Macmillan, London, 3–20

# 6 The interaction of small peptides with the various cellular interfaces of the blood–brain barrier

The neurons, glia and brain interstitial fluid (ISF) are separated from both the blood and CSF by several cell layers of different types (Figure 6.1). As we have stated, the most direct and the shortest path length between the blood and the brain is across the endothelium of the cerebral capillaries. Although these endothelial cells are joined by a continuous band of tight junctions which constitute the blood–brain barrier, the large surface area of this interface, some 5000 times greater than any other, will dominate all transport, receptor and metabolic processes between the blood and the brain[1–3].

Theoretically, any neuropeptide interaction with the blood–brain barrier may involve processes which can be located at: (a) the luminal or blood interface, (b) the abluminal or ISF interface, (c) the glial-end foot processes' surface which is in apposition with the abluminal side of the brain capillary wall, and (d) the peptidergic nerve terminals juxtaposed, to the brain microvessels. Similarly, interactions of neuropeptides with the blood–CSF barrier may be located at: (a) the blood side of the choroid epithelium, (b) the CSF side of the choroidal epithelium, and (c) peptidergic nerve terminals found in the connective tissue of the choroid plexus stroma. However when exchanges between the CSF and brain ISF are considered, the cells of the ependyma lining the ventricles do not constitute a barrier, since the pathways between the cells are open to large macromolecules. Exchanges between these compartments will occur mainly by free diffusion and may also be combined with transport via tanycytes, cells which connect the plasma, the neural ISF and CSF[4].

In recent years several different experimental techniques have been developed to characterize transport phenomena, receptor binding and metabolic transformation during the interaction of neuropeptides with these

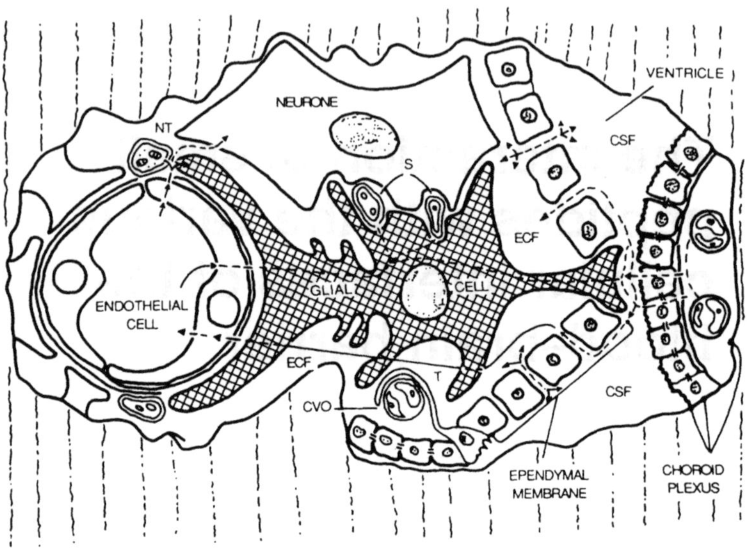


**Figure 6.1** Schematic representation of possible locations for blood-borne and/or brain-derived peptide interactions with the non-neural cellular elements in the brain (taken from Zlokovic *et al.* (1988)[1] CSF, cerebrospinal fluid; ECF, extracellular fluid of the brain; S and NT, peptidergic synapses and nerve terminals; CVO, circumventricular organs; T, tanycytes; solid arrow and lines illustrate known locations and routes for neuropeptide transfer; interrupted arrows and lines illustrate a possible fate of neuropeptide after the transfer

non-neural and the neural cellular elements of the CNS. However, none of these methods is, as yet, able to monitor in the same experiment the route of passage, receptor-mediated uptake and metabolic fate during blood–brain and blood–CSF exchanges of peptides. Thus an integrated view of neuropeptide homoeostasis within the CNS can only be obtained by combining the data from several different experimental approaches. Each of the methods usually can only yield information about one particular aspect such as metabolic transformation, receptor-mediated signal transduction or peptide compartmental transfer, and can only give results from one of the sites of possible exchange between the blood and CNS.

## PEPTIDE TRANSPORT DOGMA

Several attempts have been made to measure the first-pass extraction of small peptides into the brain during a single passage through the cerebral circulation by means of the intracarotid method of Oldendorf and/or by estimating the extraction fraction following intracarotid injection without a reference molecule. Apart from an early study with enkephalins[5], most of these short-term (up to 15 s) kinetic studies have failed to demonstrate a significant transport of peptides from blood to brain. The calculated brain-

**Table 6.1** Extraction of radiolabelled peptides by rat brain studied following intracarotid injection during short-term (5 to 15s) kinetic experiments *in vivo*. Inert polar molecules BUI values are included for comparison

| *Peptide* | *Number of amino acid residues* | *Extraction (%)* | | *Ref* |
|---|---|---|---|---|
| Carnosin | 2 | a | 1.1 | 90 |
| Gly-Phe | 2 | a | 0.9 | 91 |
| Gly-Leu | 2 | a | 0.5 | 91 |
| Glutathione | 3 | a | 0.6 | 90 |
| TRH | 3 | a | 1.0 | 90 |
| | | a,b | 0.15 | 25 |
| Enk(Leu) | 5 | a | 2.4 | 90 |
| | | a,b | none | 68 |
| DADLE | 5 | a,b | 2.0 | 68 |
| Enk(Met) | 5 | c | 15 | 5 |
| | | a | 2.4 | 90 |
| β-casomorphin-5 | 5 | d | 1.9 | 92 |
| DGAVP | 8 | d | 1.2 | 93 |
| $CCK_8$ | 8 | a | none | 6 |
| VP(Arg) | 9 | d | 2.4 | 93 |
| VP(Lys) | 9 | d | 1.3 | 93 |
| OT | 9 | d | 1.3 | 93 |
| DSIP | 9 | e | measureable | 11 |
| GnRH | 10 | a | 1.1 | 94 |
| | | d | 1.3 | 95 |
| SP | 11 | d | 0.5 | 96 |
| SRIF | 14 | d | 0.01 | 6 |
| β-E | 31 | d | 1.9 | 97 |
| *Inert polar molecules* | *Molecular weight* | | *BUI (%)* | |
| D-mannitol | 167 | | 1.8 | 6 |
| | | | 2.12 | 25 |
| Sucrose | 342 | | 2.02 | 68 |
| Inulin | 5000 | | 1.4 | 6 |
| Dextran | 70,000 | | 0.9 | 6 |

[a]Extraction was calculated by the method of Oldendorf with $^{3}$H-labelled peptide and [$^{14}$C]butanol as a reference molecule including a correction for the residual vascular radioactivity; [b]Denotes hydrolysis of peptide during single pass; [c]Extraction of reference molecule $^{3}$HOH was measured in a separate group of animals. [d]Extraction fractions were calculated without reference molecule and correction for the residual vascular radioactivity. [e]Brain/tissue plasma ratio was determined after intracarotid injection

uptake for a variety of peptides (from 2 to 31 amino acids) was in the range of that measured for the inert polar molecules, such as mannitol, sucrose, inulin and dextran (Table 6.1). These results, if taken on their face value, exclude the possibility that neuropeptides can cross the blood–brain barrier from blood to brain in significant amounts, to exert their well documented central nervous system effects. Such effects have been observed after peripheral administration of pharmacological concentrations of various peptides. Therefore the hypothesis that circulating neuropeptides cannot cross the blood–brain barrier to reach their target areas in the brain (based on the single-pass studies) requires that an alternative explanation be developed to explain the mechanism of the central action of these molecules.

Some pitfalls in the measurement of barrier permeability to neuropeptides have been enumerated by Oldendorf and others[6,7], but the absence of a significant rapid penetration of Enk(Leu), DADLE, Enk(Met), VP, DSIP, $\beta$-carnosine, TRH, SRIF, BBS, CCK and NGF has been confirmed. However, as Hugh Davson[3] pointed out, the single-pass techniques, although most valuable to study rapidly penetrating nutrients, cannot distinguish between an absolute impermeability and a slow penetration. This has been supported by experimental findings, for instance the extraction of sodium and chloride which is negligible during a single pass, yet these ions still have measurable uptake by the brain as demonstrated by the i.v. infusion method over a period of several minutes or in steady-state experiments[3,8,9]. Also it has recently been shown that DSIP estimated by radioimmunoassay may penetrate the brain[10] and that there is a significant brain penetration of [$^{125}$I]DSIP relative to RISA[11] as well as low permeability of the blood–brain barrier to nanomolar concentrations of immunoactive $\alpha$-MSH[98]. These findings oppose the view that peptides are non-penetrable substrates for the blood–brain barrier, which consequently has stimulated further research on this topic.

## PEPTIDE TRANSPORTERS AND RECEPTORS AT THE BLOOD–BRAIN BARRIER

The free movement of *circulating hydrophylic* substrates from blood to brain ISF is markedly retarded, and it has been accepted that any molecule, above a limiting size, circulating in the blood may gain access to the brain ISF only if there is a specific *transport system* for that molecule localized in the brain capillary endothelium. Such transport across the blood–brain barrier must involve three steps: (a) transport across the luminal interface, (b) transfer through the 0.3–0.5 mm thick endothelial cytoplasm, and (c) transport across the abluminal interface.

So far, specific transport systems across the blood–brain barrier have been described for several classes of nutrients which include amino acids, hexoses, monocarboxylic acids, choline, purine bases and nucleosides[12,13]. Also for thyroid hormones[14], water soluble vitamins, such as thiamine[15] and for several classes of peptides[13,16] transport or receptor-mediated transfer has been demonstrated.

The kinetic constants for the nutrients and thyroid hormones determined with the carotid artery single-injection technique are given in Table 6.2. However, it should be noted that the affinity of the blood–brain barrier for neutral amino acid transport may be even higher than previously thought, as shown recently by the vascular brain perfusion method (see Chapters 2 and 3).

**Table 6.2** Kinetic constants of the blood to brain carrier-mediated transport systems for nutrients and thyroid hormones obtained by the method of Oldendorf (according to Pardridge[13])

| *Carrier substrate* | *Representative* | $K_m$ ($\mu$M) | $V_{max}$ (nmol min$^{-1}$ g$^{-1}$) |
|---|---|---|---|
| Hexose | glucose | 11,000 ± 1,400 | 1,420 ± 140 |
| Monocarboxyic acid | lactic acid | 1,800 ± 600 | 91 ± 35 |
| Neutral amino acid | phenylalanine | 26 ± 6 | 22 ± 4 |
| Amine | choline | 340 ± 70 | 11 ± 1 |
| Basic amino acid | arginine | 40 ± 24 | 5 ± 3 |
| Nucleoside | adenosine | 25 ± 3 | 0.75 ± 0.08 |
| Purine base | adenine | 11 ± 3 | 0.50 ± 0.09 |
| Thyroid hormone | triodothyronine | 1.7 ± 0.7 | 0.19 ± 0.08 |

Transport systems for the above mentioned classes of molecules have been characterized mostly on the basis of *transport kinetic studies* by means of a variety of techniques, such as the intracarotid injection method of Oldendorf, the paired-tracer indicator dilution method of Crone, isolated brain capillaries, ventriculo-cisternal perfusion, vascular brain perfusion, and positron-emission tomography, which measure the brain extraction and the compartmental distribution of a given substrate (see Chapter 2). These studies have produced a number of transport and binding constants, which can be compared by the use of specific transforming equations[17]. The transporters as the key part of these specific transport systems at the blood–brain barrier have not as yet been isolated from the blood–brain barrier or characterized biochemically as separate distinct molecules for all classes of substrates. Nevertheless, this fact does not rule out the possibility of the existence of several specific transport systems at the blood–brain barrier, but rather emphasizes the difficulties in the full chemical purification and isolation of transport molecules. Probably, from the molecular biology viewpoint, the best characterized transport system in the blood–brain barrier is the *glucose carrier*, which has recently been photo-affinity labelled using [$^3$H]cytochalasin B[19,20] and an estimate of the molecular mass of 53 kDa for this system has been reported. However, it is of interest that it has been shown that there are tenfold more glucose transporter cytochalasin B binding sites in brain capillaries than in brain synaptosomes[20]. An attempt to identify and characterize the cerebral microvessels and the choroid plexus nucleoside transporter has been made with a similar approach using [$^3$H]nitrobenzylthionine (NBMPR) as a specific probe, and a transporter protein of approximately 60 kDa has been identified[21].

The distinction between enzymes, receptors and transporters for circulating substances seems to disappear at the blood–brain barrier with realization that receptors may also function as enzymes, and that they may internalize their ligands into the cytoplasm of the endothelial cells[2,17,22]. It is noteworthy that a membrane constituent qualifies for the predominant role of *receptor*, if the rate of ligand dissociation is greater than the rate of its internalization. In contrast, if the internalization rate is orders of magnitude faster than the dissociation rate, the protein is more properly classified as a *transporter*. Such a description, as suggested by Gjedde and Bodsh[22], renders the half saturation concentration, i.e. the Michaelis constant, $K_m$, of receptors as a function of mainly the dissociation rate, and less of the internalization rate, while the $K_m$ of a transporter reflects more the internalization turnover rate. Mathematically expressed, the Michaelis constant derived from this definition is as follows[22]:

$$K_m = \frac{k_{to} + k_{off}}{k_{on}}$$

where $k_{to}$ is the turnover rate of the internalization process, $k_{off}$ is the rate of ligand dissociation and $k_{on}$ is the rate of ligand association with the membrane protein. According to the difference between receptors and transporters, receptors should have a much lower half-saturation constant than transporters, which is typically in the nanomolar range, while transporters have a $K_m$ value often in the millimolar or micromolar range, close to the actual substrate concentration in plasma. It has also been suggested that transporter function begins with the reception of the molecule which may be the same for both transporters and receptors. However, the transporter endows the endothelium with an apparent permeability to the hydrophylic molecules which often ranges from $10^{-7}$ to $10^{-5}$ cm. s$^{-1}$.

In the case of peptides, the situation seems to be even more complicated, so far as the distinction between receptors and transporters is concerned, since most of the peptides are normally present in the plasma in the nanomolar range concentration, which is $10^6$ times lower than the concentration of nutrients. Thus, taking this factor into account, it is not surprising that short-term kinetic radiotracer studies were unable to produce much information about peptide interaction with the blood–brain barrier, since the specific activity of commercially available peptides was relatively low, so that the concentration of peptide used in these experiments was usually much above the normal level in the blood.

## LONG-TERM KINETIC STUDIES *IN VIVO*

To understand the exchange kinetics of slowly-penetrating solutes across the blood–brain barrier, it is necessary to allow test-molecules to interact with the barrier for a sufficient period of time to reveal the nature of the interaction, which is usually over a time scale in minutes[1]. An attempt to study opioid peptide penetration into the brain from the general circulation was made by i.v. injection using Ohno's method[23]. Four $^{14}C$-labelled synthetic analogues of naturally occurring opioid peptides, D-(Ala[2]) methionine enkephalinamide, $\beta$-(D-Ala[2]Homarg[9]) lipotropin 61-69, $\alpha$-(D-Ala[2]-Homarg[9])-endorphin and $\beta$-(D-Ala[2]-Homarg)-endorphin, which correspond respectively to Enk(Met), $\beta$-LPH-61-69, $\alpha$-E and $\beta$-E were studied. The permeability coefficients, estimated as parameters of the blood–brain barrier permeability for these peptides, indicated a surprisingly high penetration over periods of 3–11 minutes, which ranged from $2.5 \times 10^{-6}$ cm. $s^{-1}$ to $3.9 \times 10^{-6}$ cm. $s^{-1}$, that is comparable to molecules such as glycerol or thiourea[3]. However, for the investigation of neuropeptide transport from the periphery into the brain, this method suffers the disadvantage that it cannot distinguish between the isotopically labelled intact peptide and possible metabolically transformed fragments, which may be formed while circulating in the blood, either by plasma enzymes, or during passage through the liver and/or kidney. Even during a single passage through the brain circulation, it has been shown that

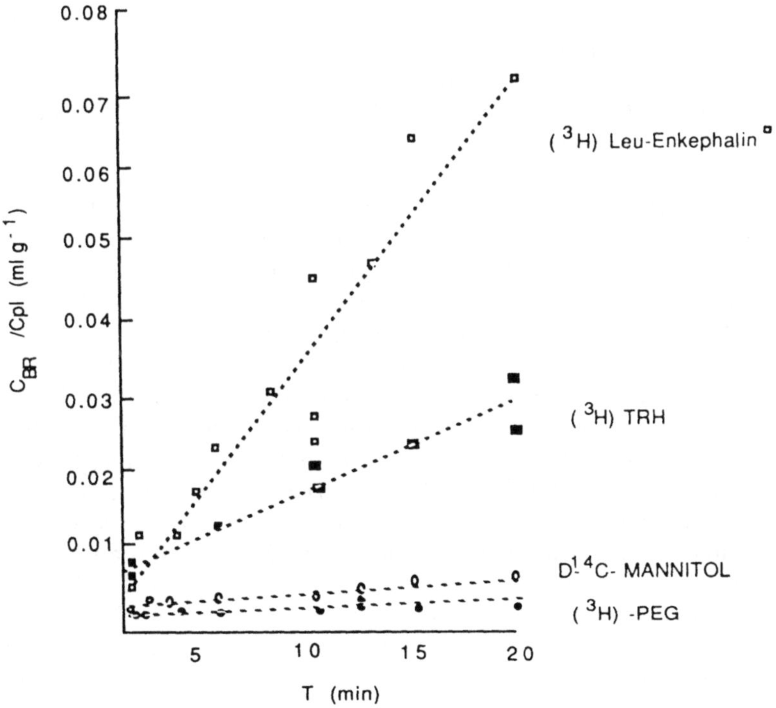


**Figure 6.2** Progressive time-dependent kinetics of cellular uptake of radiolabelled neuropeptides and inert polar molecules during vascular perfusion of the parietal cortex in the guinea-pig (taken from Zlokovic *et al.* (1987)[27] (1988)[33]

**Table 6.3** Capillary unidirectional constant ($K_{in}$), initial volume at distribution ($V_i$ ) and cerebrovascular permeability coefficient ($P$) of small radiolabelled peptides and inert polar molecules in the vascularly perfused hippocampus in the guinea-pig (according to Zlokovic *et al.* (1987)[27], (1988)[33], (1989)[30]) and unpublished observations

| *Peptide* | $K_{in}$ (ml min$^{-1}$ g$^{-1}$ x10$^{-3}$) | $V_i$ (ml/100 g) | $P$ (cm/s x10$^{-7}$) |
|---|---|---|---|
| Enk(Leu) | 3.63 ± 0.25 | 0.39 ± 0.06 | 6.05 ± 0.04 |
| DSIP | 1.61 ± 0.26 | 0.95 ± 0.24 | 2.68 ± 0.43 |
| Cyclosporin | 1.23 ± 0.27 | 0.80 ± 0.37 | 2.05 ± 0.45 |
| TRH | 1.14 ± 0.06 | 0.76 ± 0.10 | 1.93 ± 0.01 |
| *Inert molecules* | | | |
| D-mannitol | 0.22 ± 0.02 | 0.34 ± 0.07 | 0.36 ± 0.03 |
| Polyethylene-glycol (mw 4000) | 0.004 ± 0.005 | 0.30 ± 0.03 | 0.007 ± 0.008 |

the contact time of 5–15 s of the radioactive bolus with the plasma, is sufficient for the plasma peptidase activity to cleave the N-terminal amino acid residue, L-tyrosyl from both Enk(Leu) and its synthetic analogue, DADLE[24]. The same was true for TRH, when N-terminal prolin-amide has been hydrolysed from tripeptide during a single passage through the brain circulation[25]. Such a fast hydrolysis of peptides when passing as a bolus through the cerebral circulation is probably due to some 7% of the bolus mixing with the plasma[12]. This hydrolytic degradation by plasma enzymes can then yield sufficient amounts of the labelled N-terminal amino acid residue to elevate the apparent BUI values for peptides to a significant level[7].

The method of vascular perfusion of the ipsilateral forebrain in the guinea-pig[26] protects the test-molecule from systemic metabolism, and provides an adequate period of exposure of the neuropeptide to the blood–brain barrier. The method can be used to provide information with respect to the transport mechanisms across the blood–brain barrier, and the metabolic transformation during neuropeptide interactions at the blood–CNS interface. So far this method has been used to study transport, metabolic and receptor interactions at the blood–brain barrier for [$^3$H]Enk(Leu), [$^3$H]TRH, [$^3$H]VP(Arg), [$^{125}$I]DSIP and [$^3$H]cyclosporin. In contrast to the earlier studies, it has been demonstrated that there is a progressive time-dependent linear cellular uptake for various peptides into all brain regions so far investigated. For example, the uptake of Enk(Leu) and TRH has been shown into the parietal cortex, hippocampus and caudate nucleus after a 20 min exposure of the neuropeptide to the barrier[1] (Figure 6.2). Neuropeptide clearance in studied regions was several orders of magnitude greater than the brain uptake of the inert polar molecules, such as D-mannitol, polyethylene glycol or dextran. The graphically determined kinetic parameters, the unidirectional transfer rate, the initial volume of distribution or clearance constant and the $K_{in}$ and $V_i$ for tracer neuropeptides are given in Table 6.3. These results unequivocally suggest marked differences between unidirectional transfer rates both between neuropeptides as well as in comparison to the inert polar molecules.

$K_{in}$ value can provide a direct estimate of the permeability cerebrovascular constant $P$, which then can be compared with other test-molecules studied by different methods[17]. When experimentally determined $K_{in}$ values are transformed to cerebrovascular permeability coefficient, $P$, (on the assumption of the permeability surface area (PS product) of 100 $cm^2$ per gram of the brain tissue[4]) and these values are plotted against molecular weight and olive oil–water partition (P-C) for neuropeptides and inert polar molecules, no correlation was found. This analysis suggests that physicochemical factors generally do not play a major role during neuropeptide interactions with the blood–CNS barrier, although there may be exceptions.

To examine whether the estimated cerebrovascular permeability for Enk(Leu) of about $6 \times 10^{-7}$cm. $s^{-1}$ represents the true permeability for the intact peptide molecule, or the penetration of a liberated hydrolytic product (alone or together with peptide), various unlabelled competitors and inhibitors were added to the brain perfusate in different concentrations with an aim (a) to saturate the amino acid carrier systems at the barrier, (b) to inhibit the amino-peptidase activity, or (c) saturate the potential opioid receptors[27] (Table 6.4). These results have shown that neither L-tyrosine nor bacitracin exhibit any effect on the slope of [$^3$H]Enk(Leu) entry into the brain, suggesting there is no hydrolysis of peptide at the blood–brain barrier as well as no detectable transfer of peptide by the L-amino acid transport carrier. In contrast, it was possible to demonstrate a marked self-inhibition with unlabelled peptide, which suggests that the observed permeability for Enk(Leu) represents the presence of a specific saturable carrier-mediated mechanism. The opioid receptor antagonist, naloxone, does not affect kinetics of entry of Enk(Leu) into the brain, indicating that opioid receptor may not be involved in peptide interaction with the barrier. However, passive co-operativity of receptors has been investigated in more detail with new and more selective opioid receptor antagonists, and experiments with more effective $\mu$-receptor antagonists and $\delta$-receptor agonists have confirmed findings with naloxone.[29]. It has also been observed that the tetrapeptide, gly-gly-phe-leu, without the N-terminal tyrosine, had no effect on transport of Enk(Leu), whereas the tetrapeptide tyr-gly-gly-phe, without the C-terminal leucine of Enk(Leu) caused a moderate inhibition. The N-terminal part of the peptide would appear to be more important for the transport process than the C-terminal part. This finding is consistent with the view that a free C-terminal carboxyl group of Enk(Leu) is important for its activity at the $\delta$-binding site[28]. Subsequent studies have characterized Michaelis–Menten kinetic constants for the Enk(Leu) transport system in different regions of the barrier. The kinetic constants were similar to those found previously for nucleoside and purine base transport systems[29]. It was possible also to demonstrate during

**Table 6.4** Effects of various unlabelled competitors, inhibitors and receptor-antagonists on the cerebrovascular permeability constant, *P*, for small neuropeptides estimated in the parietal cortex, $^{**}p < 0.001$; $^{*}p < 0.05$

| *Peptide* | *Ref.* | *P* (cm/s x$10^{-7}$) |
|---|---|---|
| [$^3$H]Enk(Leu) | 27 | 6.03 ± 0.02 |
| +2 mM bacitracin | | 6.28 ± 0.06 |
| +5 nM L-tyrosine | | 5.92 ± 0.05 |
| +2 mM Enk(Leu) | | 1.58 ± 0.03** |
| +5 mM Tyr-Gly-Gly-Phe | | 4.27 ± 0.12* |
| +5 mM Gly-Gly-Phe-Leu | | 6.03 ± 0.04 |
| +3 mM Naloxone | | 6.15 ± 0.05 |
| [$^3$H]TRH | 33 | 2.03 ± 0.01 |
| +2 mM bacitracin | | 1.05 ± 0.01** |
| +2 mM bacitracin + 1 mM TRH | | 1.13 ± 0.01** |
| +10 mM L-prolinamide | | 1.76 ± 0.01 |
| [$^{125}$I]DSIP | 30 | 1.55 ± 0.23 |
| +7 μM DSIP | | 0.63 ± 0.25** |

these kinetic studies that there is no significant non-inhibitable component of Enk(Leu) transfer which would represent a non-saturable free-diffusion route for the peptide[29] (Table 6.5).

**Table 6.5** Kinetic constants for Enk(Leu) transport system at the blood–brain barrier (according to Zlokovic *et al.* (1989)[29])

| | $K_{in}$ (μM) | $V_{max}$ (nmol min$^{-1}$ g$^{-1}$) | $K_d$(min$^{-1}$ x$10^3$) |
|---|---|---|---|
| Parietal cortex | 39 ± 3.2 | 0.160 ± 0.022 | 0.062 ± 0.086 |
| Hippocampus | 34 ± 5.5 | 0.142 ± 0.026 | 0.068 ± 0.072 |
| Caudate nucleus | 41 ± 3.9 | 0.158 ± 0.021 | 0.070 ± 0.092 |

Experiments performed with vascular brain perfusion of $^{125}$I-labelled DSIP led to similar conclusions as the study on enkephalins (Table 6.4). Although the clearance of this peptide was somewhat lower, a progressive time dependent uptake of [$^{125}$I]DSIP gave unidirectional transfer rates significantly higher than that of the much smaller inert polar molecule D-mannitol with the cerebrovascular constant, *P*, ranging from 1.5 to 2.6 cm. $s^{-1}$ x $10^{-7}$. This clearance was almost completely self inhibited by about 7 μmol unlabelled peptide[30]. A relatively higher initial volume of peptide distribution, $V_i$, most probably reflects a binding on specific sites on the brain capillary membrane. Marked self-inhibition of the $K_{in}$ constant most likely indicates a receptor and/or transporter-mediated internalization of peptide to subsequent brain sites that are highly sensitive to L-tryptophan and may be modulated by VP(Arg). Findings with Enk(Leu) and DSIP at the blood–brain barrier parallel those demonstrated at the blood–CSF barrier for same peptides[31,32].

Vascular brain perfusion of [$^3$H]VP(Arg) has revealed a saturable mechanism at the blood–brain barrier for this nonapeptide, and a significant cerebrovascular permeability constant for the cerebral cortex, caudate nucleus and hippocampus[1], which may suggest a relatively high internalization rate. Whether endothelial vasopressin $V_1$ receptors are involved during interaction of VP with the barrier is still a matter under investigation in these *in vivo* experiments.

Studies with [$^3$H]TRH have indicated that the tripeptide may penetrate the blood–brain barrier slowly, but the finding that this uptake could be strongly inhibited by an aminopeptidase and amidase activity inhibitor, bacitracin, indicates the involvement of an enzymatic component during interaction[33]. This can be either transformation of tripeptide at the barrier occurring during the transfer from blood into the CNS, which may be accompanied by the liberation of the N-terminal amino acid residue, L-prolinamide and/or formation of the free acid (TRH-OH) (Table 6.4). The failure of unlabelled TRH to inhibit the kinetics of entry of [$^3$H]TRH in the presence of inhibitory concentrations of bacitracin, is indicative of the absence of a specific internalization mechanism for TRH at the barrier, unless an enzyme-like transport system is not present.

TRH differs from other neuropeptides in most other transport aspects observed at various locations in the CNS, as well as behaving differently with respect to its stability in body fluids[34]. Thus the rate of penetration of the intact TRH molecule, being some three times greater than that of D-mannitol, can most likely be attributed to the physicochemical factors due to its higher olive oil–water partition coefficient. It may be that the enzymatic transformation of TRH, which takes places at the blood–brain barrier, is important for the tripeptide to exert its biological potency in relation to the fact that the cyclic diketopiperazines formed after the enzymatic cleavage of TRH may be the mediators of the CNS actions of this peptide[35,36].

Studies on the kinetics of entry of the cyclic undecapeptide of fungal origin, [$^3$H]cyclosporin into the brain during vascular perfusion revealed a measurable clearance which was significantly higher than that of inert polar molecules. In a previous short-term study[38], it was shown that there is a restricted transfer of this peptide, but the method employed could not distinguish brain uptake from the distribution of the vascular space marker or from the residual radioactivity. The vascular brain perfusion method characterized the transfer of this molecule and revealed that the rate of entry is surprisingly low relative to its extremely high olive oil/saline partition coefficient. It has been suggested that relatively low cerebrovascular permeability may rather reflect the exchange of a lipoprotein–cyclosporin complex within the blood–brain barrier than the peptide itself[39].

Recently by the use of a multiple-time brain uptake analysis after the i.v. injection of $^{125}$I-labelled D-(Ala)-Peptide-T-amide, it has been shown that at the blood–brain barrier there exists a saturable transport system for this peptide[40]. This finding may be of therapeutical importance in Acquired Immune Deficiency Syndrome (AIDS), since Peptide-T from the human immunodeficiency virus envelope glycoprotein 120 has been demonstrated.

The vascular brain perfusion of neuropeptides has provided much fresh information with regard to small peptide homoeostasis in the brain. Also, a new light has been thrown on the mechanisms of interaction of peptides and the blood–brain barrier, and it has become apparent that physicochemical factors such as molecular weight and lipophilicity may only play a contributory role. The most important events underlying the mechanism by which small peptides in the blood interact with the blood–brain barrier are those which involve enzymes, transporters and/or receptors.

## BRAIN TO BLOOD TRANSPORT OF PEPTIDES

Brain to blood transport of peptides has been studied extensively by employing the ventriculo-cisternal perfusion method[37], or intraventricular injection method[41]. The ventriculo-cisternal perfusion technique, fully described by Davson *et al.*[42], uses an artificial cerebrospinal fluid containing radiolabelled peptide, which is then introduced into both lateral ventricles of the rabbit at a rate of approximately 65–70 $\mu$l min$^{-1}$, the effluent perfusate being collected from the cisterna magna under slight negative pressure. Under these conditions the perfused area for peptide interactions with CSF–CNS interfaces is represented by the lateral ventricles, the third and fourth ventricles and the cisternal space. The concentration of labelled peptide in the outflow effluent was always lower than the infused concentration, part of this being due to dilution of the perfusate by the newly formed CSF. This particular fraction can be determined from the clearance of large molecular weight Blue Dextran ($2 \times 10^6$ daltons) which is not subject to any transport and/or metabolic process in the CSF, and can be used to measure the CSF secretion rate. However, labelled peptide may be removed from the CSF perfusate by uptake into the neurons and glia across the ependymal wall, by transfer across the choroid plexus into the blood and/or by transport out of the brain extracellular space across the blood–brain barrier.

Transport of [$^3$H]Enk(Leu) as the tracer during a CSF perfusion of 140 min was studied in comparison to dextran (Figure 6.3). It was found that unlabelled tyrosine introduced into the mock CSF competitively inhibits [$^3$H]Enk(Leu) clearance, and similar effects may be obtained by adding to the perfusate, BCH (2-amino-bicyclo(2,2,1)heptane-2-carboxylic acid) the typical L-transport system amino acid substrate or the N-terminal tripeptide of Enk(Leu) tyrosyl-glycyl-glycine[37]. Since Enk(Leu) is hydrolysed by CSF to its terminal amino acid tyrosine, it is not absolutely clear from these experi-

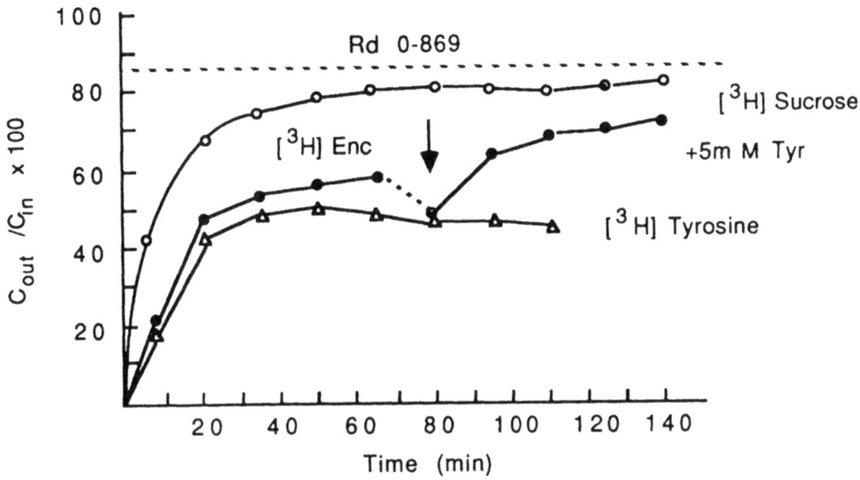


**Figure 6.3** Transport of [$^3$H]Enk(Leu) from the CSF during ventriculo-cisternal perfusion in the rabbit (taken from Begley and Chain (1982)[34]). The broken line shows the level that would be reached with no clearance and only dilution of the newly secreted CSF. At the time indicated by the arrow, the inflowing perfusate was exchanged for one containing the same activity of [$^3$H]enkephalin rises, indicating a suppression of clearance. A similar suppression of clearance is obtained with carrier-leucine and tyrosyl-glycyl-glycine at 5 mM concentrations. Clearance of [$^3$H]enkephalin was not affected by the presence of aminopeptidase and enkephalinase A and B inhibitor O-phenanthroline 1 mM of the perfusate

ments whether the observed cross-inhibitory effect between Enk(Leu) and its terminal amino acid tyrosine represents evidence for enkephalin removal from the CSF, or the self-inhibitory effect of the N-terminal amino acid residue L-tyrosine, since Enk(Leu) self-inhibitory experiments have not been carried out. DADLE, a more stable synthetic analogue, had a lower clearance than Enk(Leu) in a similar type of experiment, which would support the concept that Enk(Leu) is primarily hydrolysed in the CSF, thus the observed clearance may represent clearance of the hydrolysed N-terminal tyrosine residue, superimposed on the true Enk(Leu) clearance.

The carrier-mediated transport of enkephalins and N-Tyr-MIF-1 (a tetra-peptide with an opiate activity) has been shown to occur across the blood–brain barrier in the direction brain to blood. The intraventricular injection method was used and the measurement of the rate of disappearance of the iodinated peptide in the presence and absence of various inhibitors, was combined with the determination of the residual radioactivity left in the brain of mice after decapitation[41]. This study has characterized the peptide system which can mediate the transport of naturally occurring peptides from brain to blood, and the 'transport' system has been shown to be capable of transferring N-Tyr-MIF-1, Enk(Met), Enk(Leu) and $\beta$-casomorphin, but not kyotorphin, porcine dynorphin A, $\beta$-E and dermorphin. The study has also confirmed that an intact N-terminal tyrosine residue is a strict requirement for the system, since substitution with phenylalanine and D-tyrosine renders the compound ineffective as inhibitor, and the same applies to the blocking of the N-terminal residue. These findings demonstrate a high degree of specificity for the N-terminal end of the molecule similar to that required for the opiate activity[41]. The transport rate determined for N-Tyr-MIF-1 when the amount of radioactive peptide is kept constant while the specific radioactivity varied, or the specific radioactivity is held constant and the amount of unlabelled peptide allowed to increase, yield to the similar results, which indicate a saturable mechanism. In the case of N-Tyr-MIF-1, the entire molecule is needed for transport, since tyrosine, Tyr-Pro, Tyr-Pro-Leu and Pro-Leu-Gly are incapable of inhibiting transport of radiolabelled Tyr-MIF-1. The system is also insensitive to ouabain and theophylline, but is inhibited by furosemide and stimulated by acetozolamide. In a subsequent study, it has been demonstrated that 21 hormones, including LH, FSH, TSH, ACTH, LHRH, TRH, angiotensin I and II, GH, prolactin, thyroxine as well as several steroidal hormones, fail to inhibit the brain to blood transport system for Tyr-MIF-1 and enkephalins in mice[43].

A separate system capable of transferring vasopressin-like peptides from brain to blood has been described[44,45] using the same intraventricular injection method and measurement of the rate of disappearance of labelled peptide. This saturable system is capable of transferring some structurally similar peptides such as mesotocin, VT(Arg), pressionic amide, pressionic acid, tocinoic acid and VT(Lys), but does not transport oxytocin, VP(Lys), VP(Arg) free acid, tocinoic amide, Tyr-MIF-1 or cyclo-Leu-Glu[45].

Carrier-mediated transport from CSF has been demonstrated also for the synthetic artificial dipeptide gly-L-leu during ventriculo-cisternal perfusion in the rabbit[46]. However, it was not possible to show the same mechanism for all peptides when cleared from the CSF. Thus, for example, DSIP exhibits a clearance during ventriculo-cisternal perfusion in the rabbit which is only slightly greater than that of dextran, a marker of the ventricular space[32]. The value for the clearance of DSIP is found to be at the limit of sensitivity of the ventriculo-cisternal perfusion technique, even though sleep-peptide with high specific activity was employed. However, in contrast, this study has revealed a specific accumulation of DSIP in the hippocampus, expressed as three times increase of sucrose/DSIP R ratio. Very low clearance from CSF was also found for TRH[34], while a comparable study with $^{125}$I-labelled neurophysin, VP(Arg) and oxytocin has revealed that both nonapeptides were cleared more rapidly than their carrier neurophysin, which had a rate similar to that for inulin[48].

## INTERACTIONS AT THE BLOOD–TISSUE INTERFACE OF THE BLOOD–CSF BARRIER

Peptide interactions with the blood–tissue interface of the blood–CSF barrier were studied using the isolated perfused choroid plexus of the sheep by means of either the paired-tracer indicator dilution method with D-mannitol as an extracellular space marker, or steady state kinetic analysis with constant infusion of peptide through the choroid plexus as previously described by Segal and his colleagues for sugar transport[47]. These studies have demonstrated a significant cellular uptake of Enk(Leu) and DADLE, which can be strongly (65–76%), but not completely, self-inhibited, suggesting the presence of the saturable carrier-mediated mechanism for these opioid peptides at the blood–CSF barrier[31] (Table 6.6). Self-inhibition was significantly higher for DADLE than for Enk(Leu) whereas L-tyrosine, BCH and aminopeptidase inhibitor, bacitracin, reduced the $^{3}$H-radioactivity uptake by the choroid plexus of both enkephalins by 20 to 40%, the degree of inhibition being greater for [$^{3}$H]Enk(Leu) than for its analogue. It has been concluded that during single passage of enkephalins through the choroid plexus circulation, unidirectional uptake at the blood–tissue interface of the blood–CSF barrier consists of two components: (a) a saturable component, which represents uptake of the intact peptide by the choroid epithelium, and (b) a non-saturable component, which reflects enzymatic degradation of peptide in the blood and/or at the barrier, with a liberation of the N-terminal tyrosine residue. Higher penetrations of DADLE across the blood–CSF barrier may be attributed to its greater resistance towards hydrolysis.

Carrier-mediated cellular uptake of [$^{125}$I]DSIP has been demonstrated at the blood–tissue interface of the isolated choroid plexus by means of the steady-state method, and Michaelis–Menten type kinetics and a high affinity with very low capacity was found ($K_t$ =5.0 nM, $V_{max}$ =272 fmol. min$^{-1}$)[32]. These kinetic constants closely resemble the values found in receptor binding studies with isolated microvessels, and the possibility of receptor interaction must also be taken in account.

In an earlier study with isolated choroid plexus of the rat, accumulation of DAME in this organ was compared with that in cerebral cortical slices[49]. It has been found that the accumulation of DAME is against the concentration gradient, and may be blocked by some metabolic inhibitors, the saturable accumulation process being stronger in the choroid plexus than in the brain slices. However, although this study showed an uptake by the choroid plexus on the CSF side, the uptake may also take place on the blood side. This was confirmed in experiments where a saturable mechanism for enkephalins has been shown at the blood–tissue interface[31], as well as at the CSF side *in vivo*[37].

The permeability of the blood–CSF barrier to [$^{3}$H]TRH was studied at the blood–tissue interface of the isolated perfused choroid plexus of the sheep, using the rapid single circulation paired-tracer dilution technique, with [D-$^{14}$C]mannitol as an extracellular reference marker (see Table 6.6)[50]. The

**Table 6.6** Maximal cellular uptake ($U_{max}$) of $^3$H-labelled neuropeptides at the blood–CSF interface of the perfused choroid plexuses of the sheep. Uptake was determined by means of paired-tracer indicator dilution method of Yudilevich with D-[$^{14}$C]mannitol as an extracellular space reference carrier. Significance ** < 0.01, * < 0.05 (Zlokovic *et al.* (1986)[50], (1988)[31])

| *Peptide* | *Ref* | *Competitor or inhibitor* | *Umax (%)* | *% Inhibition* |
|---|---|---|---|---|
| [$^3$H]TRH | 50 | | 38 ± 3 | |
| | | + 10 mMTRH | 37 ± 1 | 2 |
| | | + BSA (110 g/l) | 36 ± 2 | 5 |
| | | + 10 mM Proline | 12 ± 1 | 70** |
| | | + 2 nM Bacitracin | 6 ± 2 | 85** |
| [$^3$H]Enk(Leu) | 31 | | 54 ± 4 | |
| | | + 5 mM Enk(Leu) | 19 ± 2 | 65** |
| | | + 5 mM DADLE | 26 ± 3 | 48** |
| | | + 5 mM L-Tyrosine | 35 ± 3 | 36* |
| | | + 10 mM BCH | 37 ± 4 | 32* |
| | | + 2 mM Bacitracin | 32 ± 4 | 40* |
| [$^3$H]DADLE | | | 50 ± 5 | |
| | | + 5 mM DADLE | 12 ± 2 | 76** |
| | | + 5 mM Enk(Leu) | 20 ± 2 | 61** |
| | | + 5 mM L-Tyrosine | 38 ± 3 | 24* |
| | | + 10 mM BCH | 40 ± 4 | 20* |
| | | + 2 mM Bacitracin | 36 ± 3 | 28* |

arteriovenous loss of $^{14}$C-radioactivity reflected the percentage of the D-mannitol dose which had been lost due to non-specific leakage. Cellular uptake of TRH was estimated by directly comparing venous dilution profiles of $^3$H and $^{14}$C radioactivity, and was thus independent of the degree of non-specific leakiness. The unidirectional transport of TRH could not be saturated by unlabelled tripeptide, but was markedly reduced by the inhibitor of amidase and aminopeptidase activity, bacitracin. It may be thus concluded that TRH differs from the other peptides studied with respect to the absence of a saturable mechanism and specific transport receptor system at the blood CSF barrier, but shares some similarities with enkephalins with regard to the enzymatic degradation at the barrier during a single passage through the choroid plexus circulation. Since in the presence of bacitracin there is about the same arteriovenous loss during single passage as that of D-mannitol, it is possible that tripeptide may use a non-specific free-diffusion pathway from the blood to CSF.

## RELATIONSHIP BETWEEN PLASMA AND CSF PEPTIDES

Most peptides have been identified in the CSF by various sensitive and specific radioimmunoassay procedures and the relationship between their plasma and CSF levels has often been regarded as a good indicator for peptide exchanges between their peripheral and brain pools. Early studies have shown that basal CSF and plasma levels of insulin were correlated, and that when insulin was infused in plasma, the CSF levels rose in parallel with those in the plasma[51]. Similar findings have been reported for many other peptides[52]. However, one objection of these studies is related to the uncertainty as to whether the immunoreactive material that enters the CSF remains in the intact form, or is metabolized while passing from blood into the CSF, and the radioactivity represents fragments of the metabolized peptide[53].

In later studies special attention has been focused on the above point. By the use of highly-specific antibodies that essentially require the entire peptide for cross-reactivity, it has been demonstrated that correlation occurs between CSF and plasma levels for DSIP in the basal state[53], as well as a significant increase in the CSF level after i.v. injection of [$^{125}$I]DSIP[54]. Subsequent analysis by high-affinity column chromatography has confirmed that this increase was a result of penetration by the intact molecules.

Studies investigating the relationship between plasma and CSF peptide concentrations often yield controversial results. In the case of prolactin, one study failed to find any relationship between plasma and CSF levels during human pregnancy[55], whereas another report showed a linear relationship between prolactin concentration in the two compartments[56]. A careful investigation performed in pregnant does, with analysis of the intactness of the [$^{125}$I]prolactin shown with high specific antibodies and column chromatography, has confirmed a functional dependence between plasma and CSF prolactin levels[57]. A receptor-mediated process for prolactin transport from blood to CSF as well as ready accessibility of vascular prolactin into the CSF has also been found in one study in rats, where the radiolabelled hormone was injected systemically[58].

In general, it has been accepted that plasma and CSF levels correlate when the CSF is obtained from the ventricles or the posterior fossa, but not from the area of lumbar spine. This is most likely due to the fact that CSF obtained from the lumbar area reflects spinal, rather than brain metabolism. For example, VP(Arg) displays just such a gradient and the CSF levels are equal to or less than blood levels when taken from the lumbar area, but about twice as high when taken from the cisterna magna, and about five to ten times higher when taken from the third ventricle[59]. In the case of vasopressin, it has been suggested that an elevation of peptide level on either side of the blood–brain barrier may stimulate production and release on the other, so that elevated levels in CSF do not necessarily mean that the peptide must cross the blood–brain barrier[60]. On the other hand, for some peptides, such as the opioid peptide $\beta$-endorphin, as the plasma leveles were increased a clear progressive uptake of radiolabelled peptide into the CSF has been demonstrated[61].

A number of peptides, including prolactin[62], VP(Arg)[63], thyrocalcitonin[64], $\beta$-LPH, VIP[65] and glucagon[66] were demonstrated in the CSF in relatively high concentrations, with a basal level more than 10% greater than that found in plasma. On the basis of their molecular weight and lipophilicity, assuming that blood–CSF transfer is governed only by these physicochemical factors, the estimated concentration in the CSF would be much smaller, and the expected CSF/plasma ratio would be only between 1 and 3%[12]. The higher CSF concentrations may suggest that these peptides are either selectively transported into the CSF across the blood–brain and blood–CSF barriers, or that peptides arise from the neurosecretion by brain[2]. One possible pathway for peptide neurosecretion into the ventricular space may be via transcytosis through tanacytes, which connect the portal spaces of the median eminence and the ependymal surface lining of the ventricular compartment, which has been shown by a study with apolipoprotein E[67]. Thus, receptor-mediated endocytosis of peptides at the median eminence, followed by the transfer across the tanacyte cytoplasm and exocytosis at the ependymal end of the tanacyte, may be responsible for the selective enrichment of the CSF compartment by peptides neurosecreted by the brain in the median eminence.

**Table 6.7** Concentration of adenohypophyseal hormones in human CSF (according to Lenhard and Defos (1982)[62])

| *Peptide* | *Concentration* | *CSF/blood ratio (%)* |
|---|---|---|
| ACTH | 75 pg/ml | 100 |
| CT | 250 pg/ml | 50 |
| β-E | 100 pg/ml | < 100 |
| FSH | 20 mIU/ml | 10 |
| GH | 500 pg/ml | 16 |
| LH | 25 mIU/ml | 10 |
| PRL | 1000 pg/ml | 33 |
| TSH | 2.5 mU/ml | 10 |

Adenohypophyseal peptides have also been demonstrated in the CSF in relatively high concentrations (Table 6.7), ranging from 10 to 100% of that found in plasma[62]. The simplest view would be that they have no functional significance but simply reflect hormonal blood concentrations and secretory activity of the adenohypophysis and brain. However, by circulating in the CSF these hormones may have access to specific receptors at intracranial sites, and by using this 'short loop', these hormones may have an important regulatory role either affecting the target areas of the brain and pituitary gland which are accessible via CSF, or by a feedback mechanism regulating their own secretion[3].

## ENZYMATIC DEGRADATION OF BRAIN PEPTIDES

As has been emphasized previously in this chapter, one of the major problems in the study of peptide transport and receptor interactions at different CNS–body fluid interfaces and within the brain, is whether a given peptide has been transferred from one compartment into another in an intact form, and whether the whole molecule acts on the receptor site.

Some peptides, as for instance the enkephalins, are rapidly degraded by plasma enzymes. Products of this hydrolysis, due to an aminopeptidase activity, liberate the N-terminal residue tyrosine, which may be detected even during single passage through either the cerebral[68] or the choroid plexus circulation[31]; whereas its synthetic analogue DADLE exhibits a much higher stability. Enk(Leu) is also subject to enzymatic degradation by the aminopeptidase activity of the isolated brain microvessels[69], but subsequent vascular brain perfusion studies did not reveal enkephalin hydrolysis at the blood–brain barrier from the luminal side[27], suggesting that the observed hydrolysis by the isolated capillaries may be due to activity located at the abluminal side. It has been shown that Enk(Leu) is also degraded in the CSF by a soluble aminopeptidase[37], while analogue DADLE was stable in the CSF. It is widely accepted that in the CNS the physiological actions of endogenously released enkephalins are terminated by enzymatic degradation[70,71]. Several peptidases have been implicated to participate in the brain 'enkephalinase' activity including; membrane bound neutral metallo-endopeptidase, endopeptidase 24.11, also termed as 'enkephalinase'[72,73], an extrinsic membrane associated aminopeptidase termed as aminopeptidase M II, aminoenkephalinase or puromycin-sensitive aminopeptidase[74–76], and a true integral membrane aminopeptidase called aminopeptidase M[77,78]. It has been suggested that *in vivo* neutral endopeptidase 24.11 and aminopeptidase M are responsible for the enkephalinase activity[79]. Although these peptidases play an important role in regulating the concentrations of free enkephalins in the brain, it has been clearly demonstrated that none of these are specific for the enkephalins[79]. Neutral endopeptidase 24.11 can act on different neuropeptides including substance P[80,81], NT, angiotensin I and II[82], ANF[83], chemotactic peptide[84] and bradykinin[82]. Using synthetic substrates, a broad specificity of the aminopeptidases has also been shown[75]. It is noteworthy that both aminopeptidases are not brain-specific proteins and could be found elsewhere in different tissues, in particular in the kidneys[79].

TRH is similar to ENK(Leu) so far as behaviour in plasma is concerned, where it has been cleaved rapidly with liberation of N-terminal proline amide during single passage through the brain circulation and through the choroid plexus circulation[25]. However, this tripeptide is much more stable in CSF than enkephalins[37], while the aminopeptidase and/or amidase which are most likely located at the luminal side of the blood–brain barrier, are involved in the interaction and possible transfer of TRH[33]. It has been shown that TRH may be converted to a biologically active cyclicized metabolite histidyl-proline diketopiperazin, which has been suggested to be more potent than the tripeptide in regulating some physiological processes[36]. Some other hypothalamic regulatory hormones, such as LHRH and SRIF are similarly rapidly degraded by plasma, brain tissue, gastrointestinal tract and kidney[85].

In general, we may point out that enzymatic degradation plays an important role in controlling peptide actions by initially regulating the amount of peptide available for neurosecretion at its site of production, or the amount of peptide reaching the receptor site of action. In addition, the enzymatic capacity may be involved in the control of the duration of peptide action either by inactivation at the receptor site in a target brain area, or by removal of peptide from the general circulation to prevent too large a central response from the peripheral pool. A recent study on immunohistochemical localization of aminopeptidase M in the brain has demonstrated that enzyme in the brain is localized exclusively on blood vessels[75]. Such localization requires that synaptically-released enkephalins may diffuse away from the site of release prior to degradation. Thus it has been suggested that diffusion away from enkephalinergic synapses plays a key role in terminating enkephalin action. Another interesting hypothesis has been proposed based on the mechanisms of TRH transfer from blood to brain, where the possibility of an enzyme-like transport system has been suggested[33]. Some peptides appear to be more stable, such as DSIP which may be detected in the brain in the intact form in sufficient amounts even if given orally in newborn animals[86]. Relative stability in the CSF has also been demonstrated for vasopressin, oxytocin[87], insulin[88] and neuropeptide Y[89].

## REFERENCES


1. Zlokovic, B.V., Begley, D.J., Segal, M.B., Davson, H., Rakic, Lj., Lipovac, M.N., Mitrovic, D.M. and Janokov, R.M. (1988) Neuropeptide transport mechanisms in the central nervous system. In: *Peptide and Amino Acid Transport Mechanisms in the Central Nervous System*, eds. Rakic, Lj., Begley, D.J., Davson, H. and Zlokovic B.V. Macmillan London, 3–24
2. Pardridge, W.M. (1986). Blood–brain barrier: interface between internal medicine and the brain. *Ann. Intern. Med.,* **105**, 82–95
3. Davson, H., Welch, K. and Segal, M.B. (1987). *The Physiology and Pathophysiology of the Cerebrospinal fluid.* Churchill Livingstone, London
4. Bradbury, M.W.B. (1979). *The Concept of a Blood–Brain Barrier.* John Wiley, Chichester, UK, 153–165
5. Kastin, A.J., Nissan, C., Schally, A.V. and Coy, D.H. (1976). Blood–brain barrier, half time disappearance, and brain distribution for labeled enkephalin and a potent analog. *Brain Res. Bull*, 583–589
6. Oldendorf, W.M. (1981). Blood–brain barrier permability to peptides, pitfalls in measurement. *Peptides*, **2** (suppl. 2), 109–111
7. Begley, J.D. and Zlokovic, B.V. (1986). Neuropeptides and the blood–brain barrier. In: *Blood–Brain Barrier in Health and Disease*. eds. Suckling, A.J., Rumsby, M.G. and Bradbury, M.W., Verlagsgesellschaft, FRG, 98–108

8. Abbot, J., Butt, A.M. and Zlokovic, B.V. (1988). Techniques for study of blood–brain barrier in non-mammalian species. In: *Peptides and Amino Acids in the Central Nervous System*, eds. Rakic Lj, Begley, D.J., Davson, H. and Zlokovic, B. V. Macmillan, London, 293–397
9. Sarna, G.S., Bradbury, M.W.B. and Cavanagh, J. (1978). Permeability of the blood–brain barrier after porto-caval anastomosis in the rat. *Brain Res.*, **138**, 550–554
10. Kastin, A.J., Nissen, C. and Coy, D.H. (1981). Permeability of blood–brain barrier to DSIP peptides. *Pharmac. Biochem. Behav.*, **15**, 955–959
11. Banks, W.A., Kastin, A.J. and Coy, D.H. (1984). Evidence that $I^{125}$-N-Tyr-delta sleep-inducing peptide crosses the blood–brain barrier by a non-competitive mechanism. *Brain Res.*, **301**, 201–207
12. Pardridge, W.M. (1986). Blood–brain transport of nutrients and hormones. In. *Carrier Mediated Transport of Solutes from Blood to Tissue*, eds. Yudilevich, D.L. and Mann, G.E. Pitman, London, 99–111
13. Pardridge, W.H. (1988) Recent advances in blood–brain barrier transport. *Ann. New Pharmacol. Toxicol.*, **28**, 25–39
14. Pardridge, W.M. (1979). Carrier-mediated transport of thyroid hormones through the blood–brain barrier. Primary role of albumin-bound hormone. *Endocrinology*, 105–605
15. Pratt, O.E. and Greenwood, J. (1986). Movement of vitamins across the blood–brain barrier. In: *The Blood–Brain Barrier in Health and Disease*, eds. Suckling, A.J., Rumsby, M.C. and Bradbury, M. W, B., Ellis Harwood, Chichester, 87–98
16. Pardridge, W.M. (1986). Receptor-mediated peptide transport through the blood–brain barrier. *Endocr. Rev.*, **7**, 314–330
17. Gjedde, A. (1988). Exchange diffusion of large neutral amino acids between blood and brain. In: *Peptides and Amino Acids in the Central Nervous System*, eds. Rakic, Lj., Begley, D.J., Davson H and Zlokovic, B.V. Macmillan, London, 213–223
18. Dick, A.P., Harik, S.J., Klip, A. and Wolner, D.M. (1984). Identification and characterization of the glucose transporter of the blood–brain barrier by cytochalasin B binding and immunological reactivity. *Proc. Nat. Acad. Sci. USA*, **81**, 7233–7237
19. Baldwin, S.A., Cairns, M.T., Gardinez, R.M. and Ruggirs, R. (1985). A D-glucose sensitive cytochalasin B binding component of cerebral microvessels. *J Neurochem.*, **45**, 650–652
20. Dick, A.P.K. and Harik, S.I. (1986). Distribution of the glucose transporter in the mammalian brain. *J. Neurochem.*, **46**, 1406–1411
21. Kalaria, R.W. and Harik, S.I. (1986) Nucleoside transporter of cerebral microvessels and choroid plexus. *J. Neurochem.*, **47**, 1849–1856
22. Gjedde, A. and Bodsch, W. (1987). Facilitated diffusion across the blood–brain barrier: interactions between receptors and transporters. *Wiss Z, Karl-Marx Univ. Leipzig, Math–Naturwiss. R.*, **36**, (1), 67–71
23. Rapoport, S.I., Klee, W.A., Pettigrew, K.D. and Ohno, K. (1980). Entry of opioid peptides into the central nervous system. *Science*, **207**, 84–86
24. Zlokovic, B.V., Begley, D.J. and Chain-Eliash, D.G. (1985). Blood–brain barrier permeability to leucine-enkephalin, D-alanine$^2$ D-leucine$^5$-enkephalin and their N-terminal amino acid (tyrosine). *Brain Res.*, **336**, 125–132
25. Zlokovic, B.V., Segal, M.B., Begley, D.J., Davson, D.J. and Rakic, Lj. (1985). Permeability of the blood–cerebrospinal fluid and blood–brain barriers to thyrotropin releasing hormone. *Brain Res*, **358**, 191–199
26. Zlokovic, B.V., Begley, D.J., Djuricic, B.M. and Mitrovic, D.M. (1986). Measurement of solute transport across the blood–brain barrier in the perfused guinea pig brain: method and application to *N*-methyl-alpha-aminoisobutyric acid. *J. Neurochem.*, **46**, 1444–1459
27. Zlokovic, B.V., Lipovac, N.M., Begley, D.J., Davson, H. and Rakic Lj. (1987). Transport of leucine-enkephalin across the blood–brain barrier in the perfused guinea pig brain. *J. Neurochem.*, **49**, 310–315

28. Kosterlitz, H.W., Lord, J.A.H., Paterson, S.J. and Waterfield, A. A (1980). Effects of changes in the structure of enkephalins and of narcotic analgesic drugs on their interactions with receptors. *Br. J. Pharmacol.*, **68**, 333–342
29. Zlokovic, B.V., Mackic, J.B., Duricic, B. and Davson, H. (1989) Kinetic analysis of leucine-enkephalin cellular uptake by the blood–brain barrier of an *in situ* perfused guinea-pig brain. *J. Neurochem.*, **53**, 1333–1340
30. Zlokovic, B.V., Susic, V.T., Davson, H., Begley, D.J., Jankov, R. M, Mitrovic, D.M. and Lipovac, M.N. (1989). Saturable mechanisms for delta-sleep inducing peptide (DSIP) at the blood–brain barrier of the vascularly perfused guinea-pit brain. *Peptides*, **10**, 249–254
31. Zlokovic, B.V., Segal, M.B., Davson, H. and Mitrovic, D.M. (1988). Unidirectional uptake of enkephalins at the blood–tissue interface of blood–cerebrospinal fluid barrier: a saturable mechanism. *Regul. Peptides*, **20**, 33–45
32. Zlokovic, B.V., Segal, M.B., Davson, H. and Jankov, R.M. (1988). Passage of delta sleep-inducing peptide (DSIP) across the blood–cerebrospinal fluid barrier. *Peptides*, **9**, 533–538
33. Zlokovic, B.V., Lipovac, M.N., Begley, D.J., Davson, H. and Rakic Lj (1988). Slow penetration of thyrotropin releasing hormone across the blood–brain barrier of *in situ* perfused guinea-pig brain. *J. Neurochem.*, **51**, 252–257
34. Begley, D.J. and Chain, D.G. (1982). Clearance of glutamic acid, glutamine and pyroglutamic acid from the cerebrospinal fluid of the rabbit: a comparison with thyrotropin releasing hormone. *J. Physiol.*, **326**, 22–23
35. Hersh, L.B. and McKelvey, J.F. (1979). Enzymes involved in degradation of thyrotropin releasing hormone (TRH) and luteinizing hormone-releasing hormone (LH-RH) in bovine brain. *Brain Res.*, **16**, 553–564
36. Jackson, I.M. and Lechan, R.M. (1983) Thyrotropic releasing hormone. In: *Brain Peptides*, eds. Kreiger, D.T., Browstein, J. and Martin, J. B., J. Lilly & Sons, New York
37. Begley, D.J. and Chain, D.G. (1988). Transport of encephalin from cerebrospinal fluid of the rabbit. In: *Peptide and Amino Acid Transport Mechanism in the Central Nervous System*, eds. Rakic, Lj., Begley, D.G., Davson, H. and Zlokovic, B. V. Macmillan, 55–64
38. Cefalu, W.T. and Pardridge, W.M. (1985) Restrictive transport of a lipid-soluble peptide (cyclosporin) through the blood–barrier barrier. *J. Neurochem.*, **45**, 1954–1956
39. Begley, D.J., Squires, L.K., Zlokovic, B.V. and Mitrovic, D.M. (1990). Permeability of blood–brain barrier to the immuno-suppressive cyclic peptide cyclosporin A, *J. Neurochem.*, in press
40. Barrera, C.M., Kastin, A.J. and Banks, W.A. (1987). D-(Ala)-peptide-T-amide is transported from blood to brain by a saturable system. *Brain Res. Bull.*, **6**, 629–633
41. Banks, W.A., Kastin, A.J., Fishman, A.J., Coy, D.H. and Strauss, S. L (1986). Carrier mediated transport of enkephalins and N-Tyr-MIF-1 across the blood–brain barrier. *Am. J. Physiol.*, **251**, (Endocrinol. Metab. 14), E477–E482
42. Davson, H., Hollingsworth, J.G., Carey, M.B. and Fenstermacher J, D. (1982). Ventriculo-sisternal perfusion of twelve amino acids in the rabbit. *J. Neurobiol.*, **13**, 293–318
43. Banks, W.A. and Kastin, A.J. (1988). Twenty-one hormones fail to inhibit the brain to blood transport system for Tyr-MIF-1 and the enkephalins in mice. *J. Pharm. Pharmacol.*, **40**, 289–291
44. Michals, E.A., Kastin, A.J. and Banks, W.A. (1986). Carrier-mediated transport of arginine vasopressin across the blood–brain barrier. *Clin. Res.*, **34**, 199a
45. Banks, W.A., Kastin, A.J., Horvath, A. and Michals, E.A. (1987). Carrier-mediated transport of vasopressin across the blood–brain barrier of the mouse. *J. Neurosci. Res.*, **18**, 326–332
46. Begley, D.J., Michaelson, I.A. and Davson, H. (1980). Clearance of the dipeptide glycyl-L-leucine from rabbit cerebrospinal fluid. *J. Physiol.*, **307**, 83P
47. Deane, R. and Segal, M.B. (1985). The transport of sugars across the perfused choroid plexus of the sheep. *J. Physiol.*, **362**, 245–260

48. Jones, P.M. and Robinson, I.C.A.F. (1982). Differential clearance of neurophysin and neurohypophyseal peptides from cerebrospinal fluid in conscious guinea-pigs. *Neuroendocrinology*, **34**, 297–302
49. Huang, J.T. and Lajtha, A. (1978). The accumulations of $^3$H-enkephalinamide (2-D-alanine-5-methioninamide) in rat brain tissues. *Neuropharmacol.*, **17**, 1075–1079
50. Zlokovic, B.V., Segal, M.B. and Begley, D.J. (1986). Permeability of the isolated choroid plexus of the sheep to thyrotropin-releasing hormone. In: *Carrier-mediated Transport of Solutes from Blood to Tissue*, eds. Yudilevich, D.L. and Mann, G.E. Longman, 301–306
51. Woods, S.C. and Porte, D. Jr (1977). Relationship between plasma and cerebrospinal fluid insulin levels of dogs. *Am. J. Physiol.*, **233**, E331–E334
52. Meisenberg, G. and Simmons, W.H. (1983). Peptides and the blood–brain barrier. *Life Sci.*, **32**, 2611–2633
53. Banks, W.A. and Kastin, A.J. (1985). Permeability of the blood–brain barrier to neuropeptides: the case for penetration. *Psychoneuroendocrinology*, **10**, 385–390
54. Banks, W.A. and Kastin, A.J. (1983). CSF-plasma relationships for DSIP and some other neuropeptides. *Pharmacol. Biochem. Behav.*, **19**, 1037–1040
55. Posner, B.I., van Houten, M., Patel, B. and Walsh, R.J. (1983). Characterization of lactogen binding sites in choroid plexus. *Exp. Brain*, **49**, 300–306
56. Dubney, A.K., Herbert, J., Martensz, N.D., Beckford, U. and Jones M, T. (1983). Differential penetration of three anterior pituitary peptide hormones into the cerebrospinal fluid of rhesus monkeys. *Life Sci.*, **32**, 1857–1863
57. Lorenzo, A.V., Winston, K.R. and Adler, J. (1988). The uptake by choroid plexus and passage into CSF of $^{125}$I-prolactin in preterm rabbits. In: *Peptides and Amino Acids in the Central Nervous System*, eds Rakic, Lj., Begley, D.J., Davson, H. and Zlokovic B V. Macmillan, London. 69–81
58. Walsh, R.J., Slaby, F. and Posner, B.I. (1987). Prolactin transport from blood to cerebrospinal fluid: a receptor mediated process. *Wiss, Z., Karl-Marx Univ, Leipzig, Math-Naturwiss. R.*, **36(1)**, 120–121
59. Simon-Opperman, C., Gray, D., Szczepanska-Sadowska, E. and Simon, E. (1983). Vasopressin in blood and third ventricle CSF of dogs in chronic experiments. *Am. J. Physiol.*, **245**, R541–R548
60. Szczepanska-Sadowska, E., Gray, D. and Simon-Opperman, C. (1983) Vasopressin in blood and third ventricle CSF during dehydration, thirst and hemorrhage. *Am. J. Physiol.*, **245**, R549–R555
61. Houghten, R.A., Swann, R.W. and Li, C.H. (1980). Beta-endorphin: stability, clearance, behavior, and entry into the central nervous system after intravenous injection of the tritiated peptide in rats and rabbits. *Proc. Natl. Acad. Sci. USA*, **77**, 4588–4591
62. Lenhard, L. and Deftos, L.J. (1982). Adenohypophysial hormones in the CSF. *Neuroendocrinology*, **34**, 303–308
63. Jenkins, J.S., Mather, H.M. and Ang, V. (1980). Vasopressin in human cerebrospinal fluid. *J. Clin. Endocrinol. Metab.*, **50**, 364–367
64. Morimoto, S., Nishimura, J., Miyauchi, A., Takai S.I., Okada, Y., Onishi, T., Fukuo, K., Lee, S. and Kumahara, Y. (1982). Calcitonin in plasma and cerebrospinal fluid from normal subjects and patients with medullary thyroid carcinoma: possible restriction of calcitonin by blood–brain barrier. *J. Clin. Endocrinol. Metab.*, **55**, 597–596
65. Figlewicz, D.P., Laiou, F., Dipow, A., Porte, J.D. and Woods, S.C. (1987). Gastroentero-pancreatic (GEP) peptides in the central nervous system. *Ann. Rev. Physiol.*, **49**, 383–395
66. Hoosein, N.M. and Gurd, R.S. (1984). Identification of glucagon receptors in rat brain. *Proc. Natl. Acad. Sci. USA*, **84**, 4368–72
67. Boyles, J.K., Pitas, R.E., Wilson, E., Mahley, R.W. and Taylor, J.M. (1985). Apolipoprotein, E: associated with astrocytic glia of the central nervous system and with non-myelinating glia of the peripheral nervous system. *J. Clin. Invest.*, **76**, p1501

68. Zlokovic, B.V., Begley, D.J. and Chain-Eliash, D.G. (1985). Blood–brain barrier permeability to leucine-enkephalin, D-alanine$^2$-D-leucine$^5$-enkephalin and their N-terminal amino acid (Tyrosine). *Brain Res.*, **336**, 125–132
69. Pardridge, W.M. and Meitus, L.J. (1981). Enkephalins and the blood–brain barrier: studies of binding and degradation in isolated brain microvessels. *Endocrinology*, **109**, 1138–1145
70. Schwartz, J.C., Malfroy, B. and De La Baume, S. (1981). Biological inactivation of enkephalins and the role of enkephalin dipeptidyl-carboxypeptidase ('Enkephalinase') as neuropeptidase. *Life Sci.*, 29, 1715–1740
71. Hersh, L.B. (1982). Degradation of enkephalins: the search for an enkephalinase. *Mol. Cell Biochem.*, **47**, 35–43
72. Malfroy, B., Swerts, J.P., Guyon, A., Roques, B.P. and Schwartz, J. C (1978). High affinity enkephalin-degrading peptidase in brain is increased after morphine. *Nature (London)*, **276**, 523–526
73. Kerr, M.A. and Kenny, A.J. (1974). The purification and specificity of a neutral endopeptidase from rabbit kidney brush border. *Biochem. J.*, **137**, 477–488
74. Hersh, L.B. (1985). Characterization of membrane-bound aminopeptidases from rat brain: identification of the enkephalin-degrading aminopeptidase. *J. Neurochem.*, **44**, 1427–1435
75. Hersh, L.B., Alboukhair, N. and Watson, S. (1987). Immunohisto-chemical localization of aminopeptidase M in rat brain and periphery: relationship of enzyme localization and enkephalin metabolism. *Peptides*, **8**, 523–532
76. Hui, K.-S, Wang, Y.-J and Lajtha, A. (1983). Purification and characterization of an enkephalin aminopeptidase from rat brain membranes. *Biochemistry*, **22**, 1062–1067
77. Gros, C., Giros, B. and Schwartz, J.C. (1985). Identification of aminopeptidase M as an enkephalin-inactivating enzyme in rat cerebral membranes. *Biochemistry*, **24**, 2179–2185
78. Matsas, R., Stephenson, S.L., Hryszko, J., Kenny, A.J. and Turner, A.J. (1985). The metabolism of neuropeptides: phase separation of synaptic membrane preparations with triton X-114 reveals the presence of aminopeptidase N. *Biochem. J.*, **231**, 445–449
79. McLellan, S., Dyer, S.H., Rodriguez, G. and Hersh, L.B. (1988). Studies on the tissue distribution of the puromycin sensitive enkephalin degrading aminopeptidase. *J. Neurochem.*, **51**, 1552–1559
80. Matsas, R., Fulcher, I.S., Kenny, A.J. and Turner, A.J. (1983). Substance P and [Leu]-enkephalin are hydrolyzed by an enzyme in pig caudate synaptic membranes that is identical with the endopeptidase of kidney microvilli. *Proc. Natl. Acad. Sci. USA*, **80**, 3111–3115
81. Almenoff, J., Wilk, S. and Orlowski, M. (1981). Membrane bound pituitary metalloendopeptidase: apparent identity to enkephalinase. *Biochem. Biophys. Res. Commun.*, **102**, 206–214
82. Gafford, J.T., Skidgel, R.A., Erdos, E.G. and Hersh, L.B. (1983). Human kidney 'enkephalinase', a neutral metalloendopeptidase that cleaves active peptides. *Biochemistry*, **22**, 3265–3271
83. Stephenson, S.L. and Kenny, A.J. (1987). The hydrolysis of alpha-human atrial natriuretic peptide by pig kidney microvillar membranes is initiated by endopeptidase 24.11. *Biochem. J.*, **243**, 183–187
84. Connelly, J.C., Skidgel, R.A., Schulz, W.W., Johnson, A.R. and Erdos E, G. (1985). Neutral endopeptidase 24.11 in human neutrophils: cleavage of chemotactic peptide. *Proc. Natl. Acad. Sci. USA*, **82**, 8737–8741
85. Griffiths, E.C. and Kelly, J.A. (1979). Mechanisms of inactivation of hypothalamic regulatory hormones. *Mol. Cell Endocrinol.*, **14**, 3–17
86. Banks, W.A., Kastin, A.J. and Coy, D.H. (1983). Delta sleep-inducing peptide (DSIP)-like materials absorbed by the gastrointestinal tract of the neonatal rat. *Life Sci.*, **33**, 1587–1897

87. Landgraf, R. (1981). Simultaneous measurement of arginine vasopressin and oxytocin in plasma and neurohypophysis by radioimmunoassay. *Endocrinologie*, 78, 277–284
88. Baskin, D.G., Figlewicz, D.P., Woods, S.C., Porte, D. Jr and Dorsa, D.M. (1987). Insulin in the brain. *Annu. Rev. Physiol.*, **49**, 335–347
89. Berrettini, W.H., Nurnberger, J.I. and Dimaggio, D.A. (1986). Neuropeptide Y immuno-activity in human cerebrospinal fluid. *Peptides*, **7**, 455–458
90. Cornford, E.M., Braun, L.D., Crane, P.D. and Oldendorf, W.H. (1978). Blood–brain barrier restrictions of peptides and the low uptake of enkephalins. *Endocrinology*, **103**, 1297–1303
91. Zlokovic, B.V., Begley, D.J. and Chain, D.G. (1983). Blood–brain barrier permeability to di-peptides and their constituent amino acids. *Brain Res.*, **271**, 66–71
92. Ermisch, A., Ruhle II-J, Neubert, K., Hartrodt, K. and Landgraf, R. (1983). On the blood–brain barrier to peptides: [$^3$H]beta-casomorphin-5 uptake by eighteen brain regions in vivo. *J. Neurochem.*, **41**, 1229–1233
93. Ermisch, A., Barth, T., Ruhle H-J, Skopkova, J., Hibas, P. and Landgraf, R. (1985). On the blood–brain barrier in peptides: accumulation of radioactive labelled vasopressin, desGlyNH$_2$-vasopressin and oxytocin by brain regions. *Endocrinol. Exp.*, **19**, 29–37
94. Verheugen, C., Laufer, I.R., De Facio, H., Pardridge, W.B., Lu, J.K and Judd, H.L. (1983). Impermeability of the rat blood–brain barrier to exogenously administered gonadotropin-releasing hormone. *Neuroendocrinology*, **36**, 102–104
95. Ermisch, A., Ruhle H.-J., Klauschenz, E. and Kretschumar, R. (1984). On the blood–brain barrier to peptides: [$^3$H]gonadotropin-releasing hormone accumulation by eighteen regions of the rat brain and by anterior pituitary. *Exp. Clin. Endocrinol.*, **84**, 112–116
96. Landgraf, R., Klauschenz, E., Bienert M. Ermisch, A. and Oehme, P. (1983). Some observations indicating a low brain uptake of [$^3$H]Nle$^{11}$-substance P. *Pharmazie*, **38**, 108–110
97. Ermisch, A., Ruhle, H.J., Landgraf, R. and Hess, J. (1985). Blood–brain barrier and peptides. *J. Cereb. Blood Flow Metab.*, **5**, 350–358
98. Wilson, J.F. (1988). Low permeability of the blood–brain barrier to nanomolar concentrations of immunoreactive alpha-melanotropin. *Psychopharm.*, **96**, 262–266

# 7 Transport of large peptides and proteins across the blood–brain barrier

As has been described in Chapter 6, the possibility that peptides can cross the blood–brain barrier seems to be a less controversial topic than it was a few years ago. The degree of passage of small peptides across the blood–brain barrier significantly exceeds that of the vascular markers. The most common mechanisms that govern the rate at which circulating peptides gain access to the central nervous system across the barrier membranes involves (a) *saturable* or *facilitated transfer*, and (b) *transmembrane diffusion* or passive diffusion[1,2]. This has been shown to apply to Enk(Leu), VP(Arg), TRH, DSIP and some other small peptides which are transported *in* and *out* of the central nervous system, by either or both of these mechanisms. It is interesting that most of these small peptides have molecular weights below that of thyroxine, a substance known to be transported across the blood–brain barrier by a saturable mechanism[3].

The assumption that the blood–brain barrier restricts the entry of proteins, and that this also applies to peptides ignores two facts: firstly, this view was based on early experiments with horseradish peroxidase (HRP) as the circulating 'protein'[4], which might not be applicable to naturally occurring proteins, and secondly, peptides with 50–100 amino acids are smaller than proteins such as albumin, transferrin or immunoglobulins. In addition, many large peptides have active fragments or analogues the size of small peptides, with a molecular weight below or about that of thyroxine. Thus it must be made quite clear that proteins are not the appropriate model for studying the interactions of peptides with the blood–brain barrier. However, the transfer of proteins across the barrier membranes of the central nervous system deserves particular attention, as a separate category of *macromolecular transport*. In light of the clear distinction between the transport characteristics of the small peptides and that of proteins, one may assume that large peptides may share some of the features of both classes of molecules.

## INTERACTIONS WITH CEREBRAL MICCROVESSELS IN VIVO

Many peptides which have been detected by immunocytochemical and immunofluorescence studies in peptidergic neurones and their nerve terminals are found in close proximity to the cerebral blood vessels[5–7]. These findings provide a morphological basis for a possible link between neuronal peptidergic activity and cerebro- vascular function, both (a) with respect to an active involvement of the brain endothelial cell in peptide transfer into the neural microenvironment[2,8] and (b) by supporting the concept of neurogenic peptidergic regulation of cerebral blood flow and blood–brain barrier permeability[7,9,10]. Studies on isolated brain capillaries have proved to be a useful *in vitro* model system for the characterization of many different endothelial receptors for large peptides and proteins for which kinetic constants, half-saturation binding ($K_d$) and maximal binding ($B_{max}$) are given in Table 7.1. The function of these blood–brain barrier peptide receptors may be to (a) act as a transport system, (b) initiate a signal transduction pathway, or (c) to alter blood–brain barrier permeability to circulatory nutrients, water or plasma proteins[11].

Receptor studies with isolated human brain capillaries have shown that human blood–brain barrier *insulin* receptors may be readily identified on microvessels, and that the saturation of insulin binding quantitatively parallels that found in the isolated bovine capillaries[12]. Some cross-binding can be obtained by pro-insulin, while other peptides such as SRIF or glucagon were without effect. In addition to binding to the cell surface, insulin was also found in the capillary cytoplasm as non-saturable cellular binding and was resistant to a mild acid wash, which suggests the presence of an internalized insulin. This data indicated that there was a *receptor-mediated endocytosis* which was temperature sensitive. Subsequent studies have demonstrated that isolated human brain capillaries may also release unmetabolized [$^{125}$I]insulin at 0°C, and this has been confirmed by HPLC analysis of the medium $^{125}$I radioactivity at different times of exocytosis[8]. These studies strongly suggest that both endocytosis and exocytosis occur *in vitro* in the isolated capillaries. That the same is true *in vivo* has been shown by [$^{125}$I]insulin perfusion through the intracarotid artery, of anaesthetized suckling rabbits. When subsequent HPLC analysis of acid ethanol extracts of the brain were taken the majority of the brain radioactivity co-migrated with the insulin standard and was also immunoprecipitable with insulin antibodies[13]. Thaw-mount autoradiography has also identified radioactive grains in the brain parenchyma after an intracarotid infusion of [$^{125}$I]insulin, and taken together with HPLC analysis, these findings have provided *in vivo* evidence for the transport of circulating insulin into the brain via a receptor-mediated mechanism[13].

The observation, that these receptors may be a part of the large peptide transport system at the blood–brain barrier, raises the question whether the same receptors are responsible for the observed peptide effects on the central nervous system. It has been demonstrated that there is a functional

**Table 7.1** Blood–brain barrier peptide receptors–kinetic constants, $K_d$, half-saturation binding constant (dissociation constant) and $B_{max}$, maximal binding capacity (taken from Pardridge (1988)[11])

| *Species* | *Peptide* | $K_d$ (nM) | $B_{max}$ (pmol/mg protein) |
|---|---|---|---|
| Human | Insulin | 1.2 ± 0.5 | 0.17 ± 0.08 |
| | IGFI | 2.1 ± 0.4 | 0.17 ± 0.02 |
| | IGFII | 1.1 ± 0.1 | 0.21 ± 0.01 |
| | Transferrin | 5.6 ± 1.4 | 0.10 ± 0.02 |
| Bovine | Insulin | 0.44 | 0.18 |
| | IGFI | 2.0 | 1.7 |
| | IGFII | 1.8 | 1.0 |
| | ANF | 0.11 | 0.058 |
| Canine | Angiotensin II | 1.1 | 0.022 |

difference and compartmentalization between insulin receptors on neurons and glia, compared with those on the endothelial cells, and this has been supported by some structural differences. In an isotonic homogenate of brain, the microvessels quantitatively may be removed by low-speed centrifugation in the P1 pellet, and a subsequent high speed centrifugation of the supernatant results in the second pellet (P2), which represents a crude synaptosomal preparation of mixed neuronal and glial membranes. Structural analysis of the insulin receptor in the brain P2 fraction has demonstrated protein with a molecular weight of 116 kDa and lacking the necessary carbohydrate moiety that causes binding of the receptor to a wheat germ agglutinin (WGA) affinity column[14,15]. In contrast, brain capillary insulin receptors from the brain P1 fraction has a molecular weight of 127 kDa, and both the human and bovine brain insulin capillary receptors bind to a WGA affinity column[12]. This structural difference most likely represents post-translational modification in the carbohydrate moiety, and is probably responsible for important differences in subcellular functions of insulin receptors in the brain microvessels vs. neurons and glia[12].

Besides insulin receptors, in the isolated brain capillaries, receptors for the *insulin-like growth factors*, IGF I and IGF II, were demonstrated[16]. Like insulin, IGF I and IGF II are internalized by human isolated brain capillaries, the rate of internalization being higher for the IGF II. The affinity of IGF receptors is similar to that with insulin receptors for the respective ligands (Table 7.1), but the maximal binding or capacity for the IGF receptors is approximately about 5–10 times higher in bovine capillaries. The molecular weight of the alpha-subunit of the IGF I receptor is 133 kDa isolated from the bovine capillaries[11,16]. The finding of receptor-mediated transport of IGF I and IGF II suggests that brain growth and development may be strongly influenced by circulating mitogens, and that transport of blood-borne growth factors across the blood–brain barrier may represent an important regulatory step in *neuronal growth* and *development*.

The presence of the receptors for *transferrin* in the human and rat brain capillaries has been demonstrated using monoclonal antibodies to the transferrin receptor and immunoperoxidase reaction[17]. It has been demonstrated that transferrin binds in a rapid and saturable way to isolated human capillaries[18]. In the endothelial cells of the non-brain tissues, transferrin receptors could not be identified and this absence has been attributed to the fact that transferrin, the major iron transport protein in blood, does not need specific endothelial transport system in non-brain areas, since there are large non-specific leakage pathways permitting these molecules to pass into the interstitial fluid[17,18]. However, it has also been shown that mRNA encoding the transferrin molecule is present in the brain[19,20], which suggests that this peptide may be directly neurosecreted by the nerve cells, in addition to receptor-mediated transport from blood to brain. Similar findings have recently been reported for the choroid plexus[20,21].

Recent studies have also revealed that [$^{3}$H]cationized albumin is avidly taken up by a saturable process by the isolated bovine capillaries[22], although in a previous study with native [$^{3}$H]albumin it was not possible to demonstrate a significant binding[23]. These results may suggest that cationization of proteins may represent the mechanism by which these macromolecules are rapidly taken up by the blood–brain and blood–CSF barriers via a receptor-mediated transport process. This study is also in agreement with earlier work which showed a marked increase in the entry of the circulating basic albumin into the CSF[24].

In addition, a measurable transfer of immunoglobulin G (guinea-pig) during the vascular perfusion of the guinea-pig brain has been found to be significantly higher across the blood–brain barrier, compared to inert polar molecules, such as dextran or mannitol[25]. This has been demonstrated both by using $^{125}$I-labelled immunoglobulin G, and by immunocytochemical avidin–biotin staining of immunoglobulins after a 10-minute period of brain perfusion.

## PROTEIN-BOUND MOLECULES

It has been demonstrated that many lipid-soluble substrates including steroid hormones, free fatty acids and some drugs are transported across the blood–brain barrier by a lipid-mediated transport process[26], i.e. by simple diffusion across the endothelial cells. However, this can only be true for the free fraction in plasma, since these compounds are generally bound to various plasma proteins such as albumin or specific globulins. According to the *free hormone-drug* hypothesis, it has been postulated that only this free fraction is available for transport exchanges between blood and brain, as well as other organs *in vivo*, but recent studies have demonstrated that the bound fraction is also available for transport. This suggestion arises from the findings that an enhanced rate of ligand dissociation from circulating plasma proteins

**Table 7.2** Comparison of plasma protein dissociation constant, $K_d$, *in vivo* in the brain capillary and *in vitro* (taken from Pardridge (1988)[11])

| | | $K_d$(μM) | |
|---|---|---|---|
| *Plasma protein* | *Ligand* | *in vitro* | *in vivo* |
| Albumin | testosterone | 53 ± 1 | 2,520 ± 710 |
| | tryptophan | 130 ± 30 | 1,670 ± 110 |
| | corticosterone | 260 ± 10 | 1,330 ± 90 |
| | dihydrotestosterone | 53 ± 6 | 830 ± 140 |
| | estradiol | 23 ± 1 | 710 ± 100 |
| | propranolol | 290 ± 30 | 220 ± 40 |
| | bupivacaine | 141 ± 10 | 211 ± 107 |
| | triiodothyronine | 4.7 ± 0.1 | 46 ± 4 |
| Orosomucoid | propranolol | 3.3 ± 0.1 | 19 ± 4 |
| | bupivacaine | 6.5 ± 0.5 | 17 ± 4 |

takes place *in vivo*, most likely due to conformational changes of the ligand binding site. These conformational changes are caused during transient interaction between the plasma-carrier proteins and the endothelial cells. The dissociation constants, $K_d$, determined for various plasma-protein–ligand complexes *in vitro* and *in vivo* are given in Table 7.2. From this table it can be seen that steroid hormones, tryptophan and triiodothyronine, bound to albumin, as well as amines bound to orosomucoid, have a several-fold greater dissociation constant *in vivo* than *in vitro*. This has led to the suggestion that plasma carrier proteins may actually deliver ligands to the brain via endothelial-induced conformational changes of the ligand binding within the microcirculation[11]. For example, corticosterone is readily exchangeable in brain capillaries *in vivo* from the circulating-albumin bound and corticosteroid binding globulin (CBG) pools[26]. By using human [$^3$H]CBG, it has been demonstrated that there is a negligible involvement of specific CBG receptors on brain capillaries in this process, since significant transport or specific binding of this plasma protein could not be demonstrated. On the other hand, the developmental changes in the brain uptake of circulating testosterone and of testosterone-binding proteins, such as testosterone-binding globulin (TeBG) or albumin, may play a role in the sexually dimorphic changes in brain structure that are mediated by circulating testosterone[26]. It has been demonstrated that the binding systems for testosterone are modulated in a parallel fashion in both the serum and brain compartments, and that the uptake mechanism for serum binding proteins such as TeBG and, to a lesser extent, albumin exist in the capillaries of developing rabbits.

## RECEPTOR-MEDIATED TRANSCYTOSIS AT THE BLOOD–BRAIN BARRIER

There is now enough experimental evidence derived both from *in vivo* and *in vitro* studies to accept the hypothesis that the transfer of some large peptides and proteins across the blood–brain barrier is *receptor-mediated*. This receptor-mediated transport of peptides and proteins across the blood–brain barrier is an important component in the homoeostasis of brain peptides and proteins. The model of the blood–brain barrier peptide and protein transcytosis proposed by Pardridge[11] is analogous to the *transcytosis model* of Bar *et al.*[27] for the transfer for macromolecules across capillary endothelial barriers outside the CNS. According to this model, the process of transcytosis involves three major steps: (a) *endocytosis* at the luminal membrane, or blood side of the blood–brain barrier, (b) diffusion across the endothelial cytoplasm, presumably in non-clathrin containing smooth vesicles and c) exocytosis at the abluminal membrane, or brain side of the blood–brain barrier[8,28] (Figure 7.1). It is noteworthy, the distinction between the *receptors* as a part of the peptide transport system, and *peptide carriers* is difficult to make[28], and probably in some cases may represent the same proteins. On the other hand, there is both structural and functional difference between peptide-protein *endothelial receptors* and their equivalent *neural and glial receptors* [8,15]. Also, the involvement of the peptide receptor in the process of transcytosis does not rule out the possibility that the peptide may start the transduction signal in the capillary endothelium, causing a cellular-mediated effect. Whether both transcytosis and other mediated effects occur via the same or different receptors is still open to question[2].

The above-mentioned kinetic studies with large peptides and proteins performed, either *in vitro* or *in vivo*, have provided a considerable body of evidence in favour of extensive macromolecular exchanges between blood and brain on the one hand, and blood and CSF on the other, both of which are most likely to be bi-directional. Thus the concept of a blood–brain barrier for proteins and peptides, in terms of a purely physical restrictive barrier, has to be relegated to history and is simply a product of the classic ultrastructural work with the plant enzyme HRP[4].

## STRUCTURAL ASPECTS OF THE ENTRY OF INTACT PROTEINS INTO THE BRAIN AND CSF

As discussed in Chapter 1 the term the *blood–brain barrier* was derived from the early experiments at the beginning of this century, when it was seen that certain dyes injected intravenously into the bloodstream stained most organs except the brain and CSF[29,30]. It was later demonstrated that these dyes were bound to plasma proteins and so the barrier for the dyes is actually a restriction for the dye–protein complexes. These results lead to the view that

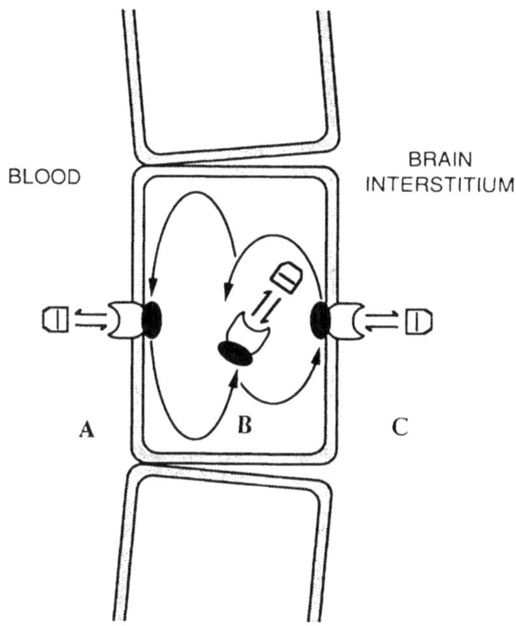


**Figure 7.1** Receptor-mediated transcytosis model for large peptide and protein transfer across the blood–brain barrier (according to Pardridge (1986)[8]). (A) receptor-mediated endocytosis, (B) movement through the endothelial cytoplasm, and (C) receptor-mediated exocytosis

proteins do not cross the blood–brain barrier, and this was further supported by ultrastructural studies with plant enzyme, HRP. This protein macromolecule, when injected intravenously, was shown, by transmission electronmicroscopy, to be stopped by the interendothelial tight junctions in the cerebral microvasculature or between interepithelial tight junctions in the choroid plexus[4,31]. These findings, for a while, completely ruled out any possibility that intact proteins could enter from blood into the brain and CSF, although a variety of proteins had been found to be present in the CSF. On the other hand, controversial data had been obtained with regard to the possibility that there is transcellular transport of intact proteins across the blood–brain and blood–CSF barriers[32]. In addition, later studies have demonstrated the transendothelial transfer of HRP from blood to brain in some segments of the fine arterioles[33] as well as from brain to blood across the cerebral capillaries[34]. However, in these studies very little attention was paid to the volumes of solution used, the concentration of protein present, or to the physiological condition of the animals in which the experiments were carried out[35–37]. For example, the absence of pinocytic vesicles or other possible transcellular pathways in most regions of the brain for HRP was based on experiments involving the injection of HRP, often in high concentrations. Little or no account has been taken of the special properties of HRP, such as stimulation of endocytosis, so the movement of HRP across the blood–brain barrier may not be the same as that for naturally occurring proteins[38].

Recently, by employing more sensitive techniques combining electron-microscopy with immunolabelling and ultra-thin cryosections with IgG gold[39], it was possible to demonstrate that both albumin and transferrin in the mature rat brain, and fetuin in the fetal sheep brain, can cross from blood into the perivascular space. Furthermore, gold particles were not seen within the endothelial tight junctions, but were found in vesicular and tubular profiles within the endothelial cells, suggesting a transcellular transport, i.e. *transcytosis*, or transport in *multivesicular bodies*[38]. Similar electronmicroscopy studies at the blood–CSF barrier of the choroid plexus of the adult rat have suggested that both cellular protein synthesis and cellular uptake followed by transcytosis can occur[38]. Thus in the epithelial cells of the choroid plexus the transcytotic pathway for transferrin[40] seems to be supplemented by the local synthesis of this protein, which has been confirmed by labelling of both endosomal/lysosomal apparatus and the protein synthetic machinery following transferrin labelling by immunogold technique. In conclusion, taking together the structural evidence and kinetic binding studies, one may say that the mechanism of macromolecular protein and large peptide transport across the blood–brain and blood–CSF barrier most likely involves receptor-mediated endoexocytosis. This is a highly specific process, and very different from the non-specific fluid phase endocytosis, by which HRP is probably taken up by the endothelial cells.

## REGULATION OF PROTEIN TRANSPORT ACROSS THE BLOOD–BRAIN BARRIER

For a long time there has been no general agreement with respect to the mechanism of the macromolecular transport across the blood–brain barrier; the main controversy centres on the respective roles of the pinocytotic transport vs. opening of the interendothelial tight junctions[41]. On the other hand, in several pathological conditions, including arterial hypertension, cerebral ischaemia, seizures, Alzheimer's disease or brain tumours, the role of extravasated albumin in the formation of brain oedema and passage of macromolecules from blood into the brain has been well documented. Recent studies have emphasized that under various pathological circumstances, apart from the mechanical component in opening the blood–brain barrier, an important role can be attributed to the enhanced vesicular transendothelial activity, rather than to an enlargement of the path through the tight junctions between endothelial cells[42–45].

It has been demonstrated that *cyclic nucleotides*, which play a major role in regulating many of the metabolic processes in cerebral endothelial cells, are also closely involved in the regulation of the rate of macromolecular transport and pinocytosis across these cells. Thus, histamine, in a similar way to *cyclic AMP*, increases pinocytosis and macromolecular transport in isolated brain microvessels[46]. Histamine was also found to be linked to the microvasculature adenylate cyclase[47] via histamine $H_2$-receptors and enhanced the macromolecular transfer after it was perfused through the carotid artery[48]. On the other hand, during an ischaemic injury accompanied by an increase of monoamine neurotransmitter content in the brain, an increased activity of the cerebral adenylate cyclase could be demonstrated, which has been found to correlate with the formation of brain oedema accompanied by the extravasation of the Evans-blue content in the brain during the recirculatory period[49]. Apart from the fact that catecholamines, histamines and some other substances such as adenosine, some metabolites of the arachidonic acid[50] and VIP[51] may stimulate microvascular adenylate cyclase activity, it has also been suggested that some vasodilators including epinephrine[52], arachidonic acid[53] and histamine[54] can also alter the level of cytoplasmic ionized calcium via stimulation of *calcium-sensitive guanylate cyclase*, so influencing macromolecular transport. Similar results were obtained by using dibutyryl *cyclic AMP*, which has a stimulatory effect on the rate of pinocytosis and macromolecular transfer[55].

The role of hormonally-sensitive adenylate cyclase, calcium-sensitive guanylate cyclase and magnesium activated cyclic phosphodiesterase in regulating the macromolecular transfer across the blood–brain barrier via alteration of the levels of different second messenger molecules, has been well established[56]. In order to elucidate which cellular components are involved in the regulation of macromolecular transport across the cerebral endothelial cells, an attempt was made to identify the proteins in the cerebral microvessels which may be the substrates for phosphorylation, as a result of an elevated level of the second messengers[57]. It has been shown that many vasoactive substances, acting on the endothelial cells, do initiate the generation of cyclic nucleotides, which after phosphorylation of certain protein kinases, may activate the transendothelial transport. On the other hand, the calcium–calmodulin system seems to be involved in the regulation of macromolecular transfer, by slowing down the transport stimulated by the cyclic nucleotides, so preventing the phosphorylation of proteins in the microvessels. If this event occurs simultaneously with an increase of cyclic nucleotide concentration, a level of homoeostasis can be achieved.

## PROTEINS IN THE CSF

It has been established by the pioneering work of Kabat *et al.*[58] that normal CSF contains a variety of proteins, but at much lower concentrations than found in plasma. It has now been accepted that, in general, the proteins of CSF are those of the plasma, with the possible exception of some plasma lipoproteins that could not be detected in the CSF and some CSF proteins which owe their origin to brain tissue and are absent from plasma. The concentration of proteins in the CSF varies along the length of the neuroaxis, with the sex and with the age of the subject. For example, in one study of different samples of human CSF, it was found that concentrations (in mg/100 ml for total proteins) were on average 26.6 from the ventricle, 31.6 from the cisterna magna and 42 from the lumbar CSF[59], as examined in 27–127 healthy subjects. With regard to age and sex difference, Vermes[60,61] confirmed the previous findings and showed a statistically significant difference between CSF protein content of females (25.14 mg/100 ml) and males (29.7 mg/100 ml). By comparing groups of children (aged between 5 and 11 years) with the adult group (aged between 12 and 79 years), the author was able to demonstrate a significantly lower CSF concentration of proteins in the younger age groups, which was on average 20.9 mg/100 ml.

The various protein fractions, expressed as percentages of the total protein content in normal human CSF, were found respectively in men and women as follows[61]: prealbumin (5.77%; 6.78%), albumin (62.81%; 60.73%) $\alpha_1$-globulins (3.44%; 3.78%), $\alpha_2$-globulins (4.74%; 5.17%), $\beta$-globulins (9.35%; 10.06%), $\tau$-globulins (6.01%; 5.92%) and $\gamma$-globulins (7.86%; 7.30%) (withdrawn from the cisterna magna and separated by cellulose acetate electrophoresis). The $\gamma$-globulin fraction, or immunoglobulin fraction can be subdivided into different classes of immunoglobulins, the principal being immunoglobulin G. Vermes[62] has summarized the published results with respect to the concentration of different immunoglobulin fractions (mg/100 ml) for men and women, respectively, as follows: IgG (2.03; 2.41), IgA (0.11; 0.16) and IgM (0.024; 0.023).

As far as the source of proteins in the CSF in the adult is concerned, it has been agreed that the majority of proteins comes from the plasma. The classical view has attributed the entry of these macromolecules to the occasional leaks at the blood–brain and blood–CSF barriers, in contrast to the modern concept which involves the possibility of protein exchanges across the endothelial wall via transendothelial transport[38]. For some proteins, such as prealbumin (thyroxin binding prealbumin, retinol binding protein etc.), which comprise 0.5% of the total protein content in serum, but constitute some 6–10% in the CSF, it has been accepted that the molecules are formed by the brain and by the choroid plexus, followed by secretion into the CSF. Similar findings were observed for transferrin and caeruloplasmin which is synthesized by both the brain and choroid plexus epithelium[20,21,63].

## SOURCE AND TYPE OF PROTEINS IN THE CSF DURING DEVELOPMENT

High levels of proteins in the CSF in the developing brain have been found in many different species[37], the highest concentration appearing in the early stage of brain development, and the clear peak may be related most appropriately to the phase of brain maturation rather than to the time of birth (Figure 7.2). For example, it has been found that in the sheep embryo at 30 days of gestation, the total CSF protein concentration is over 1000 mg/100 ml which compares with 25 mg/ml found in the adult sheep[64]. It has been shown so far that all the main proteins identified by immunofluorescence in fetal or postnatal CSF are indistinguishable from plasma proteins[65].

On the other hand, several plasma-like proteins have been identified by immunocytochemical methods in the neocortex and early allocortex of the developing sheep[66] and there is also a body of evidence for the presence, during development, of mRNA for prealbumin, transferrin and some other proteins, in the brain or choroid plexus[21,67–69], Thus, we can say that as far as the sources of the plasma-like proteins in the developing brain are concerned, the possibility exists that both the penetration of the intact proteins across the blood–CSF barrier, and synthesis *in situ* in the brain and choroid plexuses may occur.

Some studies on the permeability of the blood–brain and blood–CSF barriers to proteins in the developing brain, based on the HRP penetration, have indicated the immaturity of the barrier system, which is reflected in the leakiness of the blood–CSF and blood–brain barrier for this protein[70,71]. These studies were criticized by others on the grounds that HRP concentrations used were enormous and that the volumes of injectate obviously af-

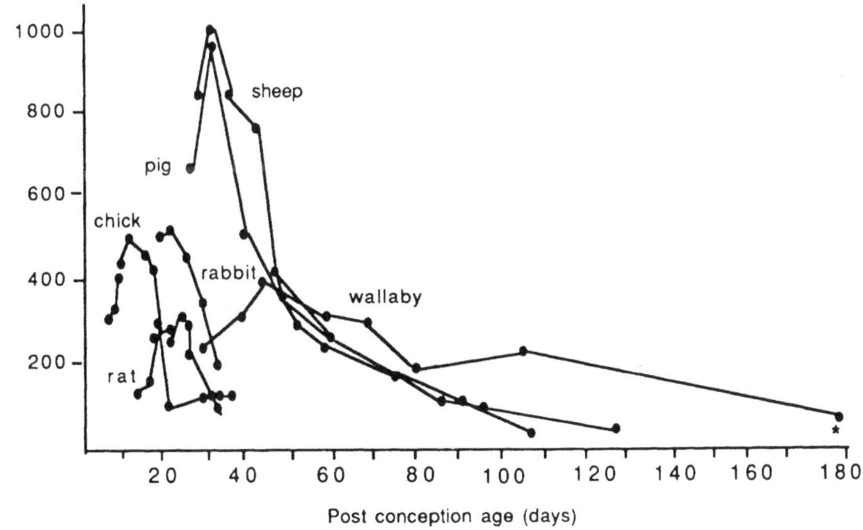


**Figure 7.2** Total protein concentration in CSF of various species during development (taken from Dziegielewska and Saunders (1988)[65])

fected the physiological condition of the fetus[65]. More careful studies, employing only trace amounts of exogenous protein followed by ultrastructural examination of the brain and choroid plexus tissues have shown that the tight junctions of the choroid plexus epithelial cells are developed very early, and that there is no correlation between the presence and complexity of the tight junctions and the concentration of protein in CSF[65].

Probably the best illustration, with respect to the characterization of the biological systems which govern protein entry into the CSF in the developing brain, derives from the steady-state experiments with different proteins. It has been shown that, for the proteins which have so far been examined, the natural steady-state (cisternal CSF/plasma ratio) may be reached in between 3 and 6 hours. The plateau area, which is representative for an established equilibrium between plasma and CSF for a given protein, is quite different for each protein, suggesting an existence of different and specific mechanisms at the blood–CSF barrier in the developing brain. For example, in the fetal sheep, the brain steady-state ratio for human albumin is about 5%, while that for the sheep's own albumin is between 10 and 15%. There is also a marked difference between proteins of similar molecular size, such as albumin and $\alpha$-fetoprotein[65].

Proteins in the fetal CSF do appear to be taken up by cells lining the ventricular system, and some evidence has been provided that this process may be both age-specific and protein specific[72–74]. The nature of the membrane specializations of these neuroependymal cells, that are responsible for the uptake of CSF proteins in the fetal brain, is described as the *CSF–brain barrier* in the fetus[38] which does not exist in the adult.

One of the possible functional significances of the high protein concentration in the developing brain may be its contribution to CSF formation by exerting a significant colloid osmotic pressure between the ventricular system and the brain extracellular fluid. This factor may be very important for the normal enlargement and shape of the growing brain[65]. On the other hand, although in the adult brain a nutritive function of the CSF seems untenable, during development and before a significant cerebral vascularization has occurred, this could be a possibility[31]. For example, a number of ligands which bind to plasma and CSF proteins are found, such as thyroxine to thyroxine-binding globulin (prealbumin), zinc to albumin and $\alpha$-macroglobulin, iron to transferrin or steroid hormones to albumin and in some species to $\alpha$-fetoprotein. It has been suggested that this exchange of ligands between CSF and plasma-carrier proteins and brain cells, may be of great importance in the different phases of brain development, before the brain becomes well vascularized[65].

## REFERENCES


1. Banks, W.A. and Kastin, A.J. (1988). Peptides and the blood–brain barrier: penetration and modulating influences. In: *Peptides and Amino Acid Transport in the Central Nervous System*, eds. Rakic, Lj., Begley, D.J., Davson, H. and Zlokovic, B. V. Macmillan, London, 21–33
2. Zlokovic, B.V., Begley, D.J., Segal, M.B., Davson, H., Rakic, Lj, Lipovac, M.N., Mitrovic, D.M. and Jankov, R.M. (1988). Neuropeptide transport mechanisms in the central nervous system. In: *Peptides and Amino Acid Transport in the Central Nervous System*, eds. Rakic, Lj, Begley, D.J., Davson, H. and Zlokovic, B. V. Macmillan, London, 3–21
3. Banks, W.A., Kastin, A.J. and Michals, E.A. (1985). Transport of thyroxine across the blood–brain barrier is directed primarily from brain to blood in the mouse. *Life Sci.*, **37**, 2407–2414
4. Reese, T.A. and Karnovsky, M.J. (1967). Fine structural localization of a blood–brain barrier to exogenous peroxidase. *J. Cell Biol.*, **34**, 207–217
5. Hendry, S.H.C., Jones, E.G. and Beinfeld, M.C. (1983). Cholecystokinin-immunoreactive neurons in rat and monkey cerebral cortex make symmetric synapses and have intimate associations with blood vessels. *Proc. Natl. Acad. Sci. USA*, **80**, 2400–2404
6. Moskowitz, M.A., Brezina, L.R. and Kuo, C. (1986). Dynorphin $\beta$-containing perivascular axons and sensory neurotransmitter mechanisms in brain blood vessels. *Cephalalgia*, **6**, 81–86
7. Owman, C., Hanks, J., Hardebo, J.E. and Kahrstrom, J. (1986). Neuropeptides and classical autonomic transmitters in the cardiovascular system: existence, coexistence, action, interaction. In: *Neural Regulation of Brain Circulation*, eds. Owmen, C., Hardebo, J.E., Elsevier Science, Amsterdam, 299–331
8. Pardridge, W.M. (1986). Receptor-mediated peptide transport through the blood–brain barrier. *Endocrine Rev.*, **7**, 314–330
9. Kobayashi, H., Magnomi, M.S., Govoni, S., Izumi, F., Wada, A. and Trabucchi, M. (1985). Neuronal control of brain microvessel function. *Experientia*, **41**, 427–558
10. Edvinsson, L., McCulloch, J., Kingman, T.A. and Uddman, R. (1986). On the functional role of the trigemino-cerebrovascular system in the regulation of cerebral circulation. In: *Neural Regulation of Brain Circulation*, eds. Owman, C. and Hardebo, J.E., Elsevier Science, Amsterdam, 407–418
11. Pardridge, W.M. (1988) Recent advances in blood–brain barrier transport. *Ann. Rev. Pharmacol. Toxicol.*, **28**, 25–39
12. Pardridge, W.M., Eisenberg, J. and Yank, J. (1985). Human blood–brain barrier insulin receptor. *J. Neurochem.*, **44**, 1771
13. Duffy, K.R., Choi, T.B. and Pardridge, W.M. (1986) *In vivo* evidence that brain insulin originates from blood. *Clin. Res.*, **34**, 57A
14. Zahniser, N.R., Goens, M.B., Hanaway, P.J. and Vinych, J.V. (1984). Characterization and regulation of insulin receptors in rat brain. *J. Neurochem.*, **42**, 1354
15. Heidenreich, K.A., Zahniser, N.R., Berhanu, P., Brandenburg, D. and Olefsky, J.M. (1983). Structural differences between insulin receptors in the brain and peripheral target tissues. *J. Biol. Chem.*, **258**, 8527
16. Frank, H.J.L., Pardridge, W.M., Morris, W.L., Rosenfeld, R.G. and Choi, T.B. (1986). Binding and internalization of insulin and insulin-like growth factors by isolated brain microvessels. *Diabetes*, **35**, 654
17. Jefferies, W.A., Brandon, M.R., Hunt, S.V., Williams, A.F., Gatter, K.C. and Mason, D.Y. (1984). Transferrin receptor on endothelium of brain capillaries. *Nature (London)*, **312**, 162
18. Pardridge, W.M., Eisenberg, J. and Yang, J. (1987). Human blood–brain barrier transferrin receptor. *J. Neurochem.*, **49**, 1394–1401

19. Levin, M.J., Tuil, D., Uzan, G., Dreyfus, J.C. and Kahn, A. (1984). Expression of the transferrin gene during development of nonhepatic tissues: high levels of transferrin mRNA in fetal muscle and adult brain. *Biochem. Biophys. Res. Commun.*, **122**, 212
20. Aldred, A.R., Dickson, P.W., Marley, P.D. and Schreiber, G. (1987). Distribution of transferrin synthesis in brain and other tissues in the rat. *J. Biol. Chem.*, **262**, 5293–5297
21. Dickson, P.W., Aldred, A.P., Marley, P.D., Guo-Fen, T., Howlett, G. J. and Schreiber, G. (1985). High prealbumin and transferrin mRNA levels in the choroid plexus of rat brain. *Biochem. Biophys. Res. Commun.*, **127**, 890–895
22. Kumagai, A.K., Eisenberg, J. and Pardridge, W.M. (1986). Rapid binding and internalization of cationized albumin by isolated brain capillaries. *Clin. Res.*, **34**, 69A
23. Pardridge, W. M. Eisenberg, J. and Cefalu, W.T. (1985). Absence of albumin receptor on brain capillaries *in vivo* or *in vitro*. *Am. J. Physiol.*, **249**, E264
24. Griffin, D.E. and Giffels, J. (1982). Study of protein characteristics that influence entry into the cerebrospinal fluid of normal mice and mice with encephalitis. *J. Clin. Invest.*, **70**, 289–293
25. Zlokovic, B.V., Skundric, D.S., Segal, M.B., Colover, J., Jankov R, H., Pejnovic, N., Lackovic, V., Mackic, J., Lipovac, M.N, Davson, H., Kasp E, Dumonde, D. and Rakic, Lj. (1989). Blood–brain barrier permeability changes during acute allergic encephelomyelitis induced in the guinea-pig. *Metab. Brain Dis.*, **4(1)**, 33–40
26. Pardridge, W.M. (1987). Plasma protein mediated transport of steroid and thyroid hormones. *Am. J. Physiol.*, **252**, E157–E164
27. Bar, R.S., DeRose, A., Sandra, A., Peacock, M.L. and Owen, W.G. (1983). Insulin binding to microvascular endothelium of intact heart: a kinetic and morphometric analysis. *Am. J. Physiol.*, **244**, E447–E543
28. Gjedde, A. and Bodsch, W. (1987). Facilitated diffusion across the blood–brain barrier: interactions between receptors and transporters. *Wiss, Z., Karl-Marx Univ, Leipzig. Math-Naturwiss. R.*, **36(1)**, 67–71
29. Goldmann, E.E. (1909) Die äussere und innere secretetion des gesunden und kranken organismus im lichte der 'vitalon tarbung'. *Beitz. Klin. Chiurg.*, **64**, 192–265
30. Goldman, E.E. (1913). Vitalfarbung am Zentral Nervensystem (Abh. Preuss. Akad. Wiss.). *Phys-Math. Kl.*, **1**, 1–60
31. Brightman, W.M. and Reese, T.S. (1969). Junctions between intimately apposed cell membranes in the vertebrate brain. *J. Cell Biol.*, **41**, 648–677
32. Balin, B.J., Broadwell, R.D., Salcman, M. and El-Kalliny, M. (1986). Avenues for entry of peripherally administered protein to the central nervous system in mouse, rat and squirrel monkey. *J. Comp. Neurol.*, **251**, 260–280
33. Westergaard, E. and Brightman, M.W. (1973). Transport of protein across normal cerebral arterioles. *J. Comp. Neurol.*, **152**, 17–44
34. van Deurs, B. (1977). Vesicular transport of horseradish peroxidase from brain to blood in segments of the cerebral microvasculature in adult mice. *Brain Res.*, **124**, 1–8
35. van Deurs, B. (1980). Structural aspects of brain barriers, with special reference to the permeability of the cerebral endothelium and choroidal epithelium. *Intern. Rev. Cytol.*, **65**, 117–191
36. Møllgard, K. and Saunders, N.R. (1986). The development of the human blood–brain and blood–CSF barriers. *Neuropath. Applied Neurobiol.*, **12**, 337–358
37. Dziegielewska, K.M. and Saunders, N.R. (1987). The development of the blood–brain barrier: proteins in fetal and neonatal CSF, their nature and origins. In: *Handbook of Human Growth and Biological Development*, eds. Meisami, E. and Timiras, P.J., CRC Press, Bora Raton
38. Møllgard, K., Balslev, Y. and Saunders, N. (1988). Structural aspects of the blood–brain and blood–CSF barriers with respect to endogenous proteins. In: *Peptides and Amino Acid Transport Mechanisms in the Central Nervous System*, eds. Rakic, Lj., Begley, D.J., Davson H and Zlokovic, B. V. Macmillan, London, 95–105

39. Møllgard, K., Balslev, Y. and Saunders, N.R. (1988)., A. combination of *in situ* synthesis and uptake accounts for transferrin-like immunoreactivity of choroid plexus epithelial cells. In: *Peptides and Amino Acid Transport Mechanisms in the Central Nervous System*, eds. Rakic, Lj., Begley, D.J., Davson, H. and Zlokovic, B. V. Macmillan, London, 93–102
40. van Deurs, B., von Bülow, F. and Møller, M. (1981). Vesicular transport of cationized ferritin by the epithelium of the rat choroid plexus. *J. Cell Biol.*, **89**, 131–139
41. Cervos-Navarro, J., Artigas, J. and Mrsulja, B.J. (1983). Morphofunctional aspects of the normal and pathological blood–brain barrier. *Acta Neuropathol. (Suppl.)*, **8**, 1–19
42. Inoue, M., Fukushima, M., Tsutsumi, K., Shibata, S., Mori, K. and Setoguchi, T. (1985). Freeze-fracture replica study of capillary endothelium after embolization in the dog. *J. Neurosurg.*, **62**, 737–742
43. Noske, W. and Hirsch, M. (1986). Morphology of tight junctions in the ciliary epithelium of rabbits during arachidonic acid-induced breakdown of the blood–aqueous barrier. *Cell Tissue Res.*, **245**, 405–412
44. Faustmann, P.M. and Dermietzel, R. (1985). Capillary junctions of the rat are not affected by osmotic opening of the blood–brain barrier. *Acta Neuropathol.*, **63**, 179–189
45. Rakic, L.M., Zlokovic, B.V., Davson, H., Segal, M.B., Begley, D.J., Lipovac, M.N. and Mitrovic, D.M. (1989). Blood–brain barrier dysfunction in the guinea-pig and chronic amphetamine intoxication. *J. Neurol. Sci.* (In press)
46. Joó, F. (1972). Effect of $N^6$, $O^6$-dibutyryl cyclic 3′,5′-adenosine monophosphate on the pinocytosis of brain capillaries in mice. *Experientia*, **28**, 1470–1471
47. Karnushina, I., Palacios, J.M., Barbin, G., Dux, E., Joó, F. and Schwartz, J.C. (1980). Studies on a capillary-rich fraction isolated from brain: histaminic components and characterization of the histamine receptors linked to adenylate cyclase. *J. Neurochem.*, **34**, 1201–1208
48. Dux, E. and Joó, F. (1982). Effects of histamine on brain capillaries: fine structural and immunohistochemical studies after intracarotid infusion. *Exp. Brain Res.*, **47**, 252–258
49. Adam, G., Joó, F., Temesvari, P., Dux, E. and Szerdahelyi, P. (1988). Effects of acute hypoxia on the adenylate cyclase activity and albumin transport of brain microvessels. *Neurochem. Int.*, in press
50. Schutz, W., Steuer, G. and Tuisl, E. (1982). Functional identification of adenylate cyclase-coupled adenosine receptors in rat brain microvessels. *Eur. Pharmac.*, **85**, 177–184
51. Huang, M. and Rorstad, O.P. (1983). Effects of vasoactive intestinal polypeptide, monoamines, prostaglandins and 2-choloroadenosine on adenylate cyclase in rat cerebral microvessels. *J. Neurochem.*, **40**, 719–726
52. Kowarski, D., Shuman, H., Somlyo, A.P. and Somlyo, A.V. (1985). Calcium release by noradrenaline from central sarcoplasmic reticulum in rabbit main pulmonary artery smooth muscle. *J. Physiol.*, **366**, 153–175
53. Wolf, B.A., Turk, J., Sherman, W.R. and McDaniel, M.L. (1986). Intracellular $Ca^{2+}$ mobilization by arachidonic acid. *J. Biol. Chem.*, **261** 3501–3511
54. Matsumoto, T., Kanaide, H., Nishimura, J., Shogakiuchi, Kobayshi, S. and Nakamura, M. (1986). Histamine activates $H_1$-receptors to induce cytosolic free calcium transients in cultured vascular smooth muscle cells from rat aorta. *Biochem. Biophys. Res. Commun.*, **135**, 172–177
55. Joó, F., Temesvari, P. and Dux, E. (1983). Regulation of the macromolecular transport in the brain microvessels: the role of cyclic GMP. *Brain Res*, **278**, 165–174
56. Joó, F. (1986). New aspects to the function of the cerebral endothelium. *Nature (London)*, **321**, 197–198
57. Joó, F. (1988). Cyclic nucleotide mediated regulation of albumin transport in brain microvessels. In: *Peptides and Amino Acid Transport Mechanism in the Central Nervous System*, eds. Rakic, Lj., Begley, D.J., Davson, H. and Zlokovic, B. V. Macmillan, London, 119–129

58. Kabat, E.A., Moore, D.H. and Landow, H. (1942). An electrophoretic study of the protein components in the cerebrospinal fluid and their relationship to serum protein. *J. Clin. Invest.*, **21**, 571–577
59. Weisner, B. and Bernkazdt, W. (1978). Protein fractions of lumbar, cisternal and ventricular cerebrospinal fluid. *J. Neurol. Sci.*, **37**, 205–214
60. Vermes, L.M.S. (1983). Proteinas do liquido cefalorraqueano. I. Estudo comparativo entre metodos de concentracao. *Arquivio do Neuro-Psychiatrica*, **41**, 108
61. Vermes, L.M.S. (1983). Proteinas do liquido cefalorraqueano. II. Valores normais das fracoes proteicas obtidas por elektroforese (variacoes ligados a cor, sexo e edade). *Arquivio do Neuro-Psychiatrica*, **41**, 9–24
62. Vermes, L.M.S. (1983). Proteinas do liquidocefalorraqueano. III. Valores normais das immunoglobulinas, G, A e M. *Arquivio do Neuro-Psychiatrica*, **41**, 25–49
63. Aldred, A.R., Grimes, A., Schrieber, G. and Mercer, J.F.B. (1987). Rat ceruloplasmen. *J. Biol. Chem.*, **262**, 2875–2878
64. Dziegielewska, K.M. (1982). Proteins in fetal CSF and plasma. PhD thesis, University of London
65. Dziegielewska, K.M. and Saunders, N.R. (1988). The origins and functions of proteins in CSF in the developing brain. In: *Peptides and Amino Acid Transport Mechanisms in the Central Nervous System*, eds. Rakic, L., Begley, D.J., Davson, H. and Zlokovic, B.V. Macmillan, London, 105–121
66. Reynolds, M.L. and Møllgard, K. (1985). The distribution of plasma proteins in the neocortex and early allocortex of the developing sheep brain. *Anat. Embryol.*, **171**, 41–60
67. Levin, M.J., Tuil, D., Uzam, G., Dreyfus, J.C. and Kahn, A. (1984). Expression of the transferrin gene during development of non-hepatic tissues: high level of transferrin mRNA in fetal muscle and adult brain. *Biochem. Biophys. Res. Commun.*, **122**, 212–217
68. Dziegielewska, K.M., Saunders, N.R. and Soreq, H. (1985). Messenger ribonucleic acid (mRNA) from developing rat cerebellum direct in vitro synthesis of plasma proteins. *Dev. Brain Res.*, **23**, 259–267
69. Dziegielewska, K.M., Saunders, N.R., Schejter, E.J., Zakut, H., Zevin-Sorkin,, D. Zisling, R. and Soreq, H. (1986). Synthesis of plasma proteins in the fetal, adult and neoplastic human brain tissue. *Dev. Biol.*, **114**, 95–104
70. Wakai, S. and Hirokawa, N. (1981). Development of blood–cerebrospinal fluid barrier to horseradish peroxidase in the avian choroidal epithelium. *Cell Tiss. Res.*, **217**, 271–278
71. Risau, W., Hallmann, R. and Albrecht, U. (1986). Differentiation-dependent expression of proteins in brain endothelium during development of the blood–brain barrier. *Dev. Biol.*, **117**, 537–545
72. Cavanagh, M.E. and Warren, A. (1985). The distribution of native albumin and foreign albumin injected in lateral ventricles of prenatal and neonatal rat forebrain. *Anat. Embryol.*, **172**, 345–351.
73. Moro, R., Fielitz, N., Esteveo, A., Grunberg, J. and Uriel, J. (1984). *In vivo* uptake of heterologous alpha-fetoprotein and serum albumin by ependymal cells of developing chick embryos. *Int. J. Dev. Neurosci.*, **2**, 143–148
74. Limbrick, A.R., Cavanagh, M.L., Reynolds, M.L. and Saunders, N.R. (1990). Uptake of intraventricular fetuin by neuroependymal cells in the developing rat brain, in press

# 8 Disease states and the blood–brain barrier related to peptides

## Brain disorders and peptides

The role of peptides in various neurological and psychiatric disorders has been extensively studied, both in clinical and fundamental research. These studies have demonstrated that neuropeptides are deeply involved in the mechanisms underlying a variety of brain disorders, including for example, a group of the affective disorders such as depression and mania, cerebral functional disorders such as schizophrenic psychosis, or extrapyramidal motor disorders such as Parkinson disease. In some of these disease states it has been possible to demonstrate a relationship between mental and endocrine functions of peptides, while in others the role of peptides has been related to their interactions with the classical neurotransmitters[1].

A number of studies were performed in the specialized clinical area of *psychoendocrinology*, where peptide levels in body fluids were determined, peptide challenge tests were performed and the possibility of therapeutical application of peptides in affective disorders was investigated. By employing the specific and highly sensitive radioimmunoassays for different peptides, their concentrations were measured in the CSF and plasma in patients with *depression* and *mania*, and either elevation, drop or unaltered peptide levels in the CSF have been demonstrated. For instance, in *depression*, the concentration in CSF of CRF, TRH and substance P have been found to be elevated, although each claim depends on a single report[2–4]. On the other hand, the concentrations of $\beta$-endorphin, BBS, DSIP, calcitonin and CCK in unipolar depression were unchanged[1,5]. The CSF concentration of SRIF, VP(Arg), CCK in bipolar depression and VIP in non-endogenous depression were reduced[6–10]. In some of these studies patients suspected of neurological disorders were used as the control group. In non-endogenous depressed patients a 50% reduction in the CSF VIP concentration relative to the control group was shown, while the plasma concentrations of this peptide were equal

**Table 8.1** Peptide level in the CSF of patients with affective disorders (according to Prange *et al.* (1987)[1])

| *Peptide* | *Depression* | *Mania* |
|---|---|---|
| CRF | ↑ | |
| TRH | ↑ | |
| SO | ↑ | |
| β-E | N | |
| CCK | N | N |
| CT | N | ↓ |
| BBS | N | N |
| DSIP | N | |
| SRIF | ↓ | N |
| VP(Arg) | ↓ | |
| VIP | ↓ (non-endogenous) | N |
| CCK | ↓ (bipolar) | N |

N = normal; ↑ increased level; ↓ decreased level

in both groups. This finding was observed before treatment, and the low level of VIP in CSF persisted after the treatment with electroconvulsive therapy or antidepressive drug therapy. According to these findings, it has been suggested that the reduced VIP level in the CSF of patients with non-endogenous depression may be either a trait-marker for this type of depression, or it is an enduring consequence of the experience[11]. In mania, the level of SRIF in CSF is found to be normal, as well as CCK, bombesin and VIP. The VP(Arg) concentration in the CSF of some manic patients was increased, while calcitonin concentration may be reduced[9,12]. The summarized findings with respect to the peptide concentrations in the CSF in patients with affective disorders, taken from different studies, are given in Table 8.1. However, although these studies have provided valuable data for the possible involvement of various peptides in the etiology of affective disorders, it still has not been made clear whether an elevated level of a given neuropeptide in the CSF represents its direct transport across the choroid plexus, transport from blood to brain and then to CSF, or, alternatively, the peptide has been released directly by the brain into the brain interstitial fluid and then into the CSF from certain neural centres in the brain. The reverse has also not yet been answered, whether the reduced level of a given neuropeptide in the CSF may represent an enhanced transport out of CSF and central nervous system, or is this rather a consequence of more restricted penetration of the peptide across the blood–brain and blood–CSF barriers, or, alternatively, it may represent its reduced release from the neural centres into the CSF.

A major effort in clinical research is now focused on the various peptide challenge tests, which are all based on a similar principle, namely to provoke secretion of the adenohypophyseal hormones by giving systemically hypothalamic releasing factors. Among first studies reported in this field were findings of diminished TSH response to TRH in euthyroid depressed patients[13–16]. It has been consistently demonstrated that about 30% of euthyroid depressed patients show a reduced TSH response. The differences in TSH response between unipolar and bipolar depressed patients, or between primary and secondary depressed patients, could not be demonstrated[17]. In depressed patients prolactin release following TRH administration was blunted[16], and the effect appears to be limited to women[18]. TRH administration may also evoke GH secretion in depressed patients[19], but this response is relatively inconsistent, and it has not been distinguished from spontaneous GH release[20]. It is interesting that manic patients not receiving lithium treatment often show, like depressed patients, a blunted TSH response to TRH[17]. However, manic patients taking lithium usually exhibit an exaggerated TSH response to TRH, probably because lithium may cause a subtle antithyroid effect. Although a body of evidence has been accumulated with respect to an altered TSH response to TRH in various affective disorders, the mechanism which underlies these changes is still a matter of controversy, and it is difficult to see how these findings relate to the conditions of mania and depression.

CRF infusion has also been used in psychiatric patients, and in those with depression, a reduced response to ACTH has been found. In manic patients no changes in ACTH or cortisol response to CRF were reported[21]. It has also been shown that in depressed patients, the ACTH response to CRF is strongly influenced by basal cortisol levels. This finding has been interpreted to mean that in depression the excess CRF activity has stimulated the pituitary–adrenal axis, resulting in hypercortisolaemia and the consequent feedback inhibition of ACTH release. LHRH infusion in depressed patients failed to show any significant changes in LH and FSH responses[22,23]. VP(Arg) and VP(Lys) were both shown to release ACTH and cortisol in healthy subjects, but in depressed patients the cortisol response to VP(Arg)[24] or to VP(Lys) is absent[25]. These findings support the hypothesis that there is a reduced cortisol and ACTH response to CRF in depressed patients.

Several peptides, peptide agonists and antagonists have been used as therapeutic agents in psychoendocrinology. TRH is certainly the most frequently used peptide as a possible therapeutic agent in depression, but its reported efficacy seems to vary according to the study. In general, oral application of TRH has no value as a remedy for depression, while i.v. administration seems to be more promising[1]. It has also been shown that in manic patients, the i.v. administration of TRH has produced encouraging results, probably via TRH central cholinomimetic actions[26,27]. So far as the antidepressant action of $\beta$-endorphin is concerned, there are more studies with positive results than with negative ones, but the use of a double-blind test has failed to reveal any difference from the effects of placebo injections[28–30]. The use of MIF-1 in depression appears to be promising even when given orally[31], but, on the other hand, no therapeutic effects of the stable Enk(Met) analogue FK-33-824 could be found in depressed patients, neither did the opioid antagonist naloxone have any effect and even may have aggravated the condition[5,32], and in manic patients the effects of naloxone were controversial.

Schizophrenia has been studied in relation to endorphins, and three main hypotheses have been proposed: (a) endorphin excess, (b) endorphin deficiency, and (c) disturbances in β-endorphin fragmentation[33]. From the clinical point of view, there is special interest in the last hypothesis, which states that a relative deficiency of the γ-endorphins which are peptide fragments derived from β-endorphin, may be concerned with pathogenesis of schizophrenic psychoses[34,35]. This hypothesis has encouraged a number of clinical trials in which the antipsychotic action of γ-endorphins were tested. These endorphins were tested using a double-blind trial in several studies, and antipsychotic action in a number of schizophrenic patients, with no side-effects, such as extrapyramidal disturbances, were obtained[36]. However, the responses of the patients did vary from no response to complete remission of psychotic symptoms. It was observed with further analysis that there is a particularly sensitive group of schizophrenic patients who, in response to γ-type endorphin treatment exhibit a variety of motor symptoms, such as retardation and catonic symptoms which inhibited the response to peptide treatment[36–38]. In addition, reduced responsiveness was observed in patients with a long duration of recent psychotic episodes and who had been treated previously with high doses of neuroleptics. Several clinical trials have also been performed to test the antipsychotic action of CCK-related peptides in schizophrenic patients, including CCK-33, CCK-8 and ceruletide[39–42]. A variety of protocols were used, such as open or single-blind, uncontrolled, double-blind, placebo-controlled and double-blind placebo cross-over[35]. In general, an improvement in symptomatology was reported for most studies with an effect, which was rather rapid and persisted for several days or weeks, even after a single injection. However, although the effect was in some cases clinically significant, the amelioration was in general partial and not always impressive. With ceruletide an effect which was prolonged and long-lasting was observed, provided the ceruletide was administered twice in two consecutive weeks[43]. However, treatment with ceruletide differs from the treatment with γ-type endorphins since ceruletide affected the positive symptoms in schizophrenia, while γ-endorphins diminished both positive and negative symptoms. The difference was also extended to the protocol of administration, since γ-type endorphins were given daily for a period of 8–10 days, whereas ceruletide was given once a week. However, the lack of significant differences between the antipsychotic action of CCK-related peptides and γ-type endorphins, is also consistent with results seen in experimental animal models of schizophrenia, which suggests that these peptides exert their effects on similar brain mechanisms. It has been suggested that future clinical trials on the similarities and dissimilarities of these classes of peptides should include investigations on the genetic factors, such as HLA antigen typing, as well as the effect of the peptide treatment on not only the psychotic symptomatology, but also on the long-term outcome of the disease[35].

Alzheimer disease has been recently investigated in relation to 4,200-dalton peptide, or amyloid ($A_4$) peptide, which has been isolated from cortical microvessels in patients with Alzheimer disease[44]. It has been shown that the amyloid peptide ($A_4$), called also Cb-peptide, represents an important component of the neuronal plaque in Alzheimer disease, which accumulates around cortical microvessels, forming the vascular amyloid. By means of a specific RIA method, using antiserum directed against a synthetic fragment of the $A_4$ peptide, it has been suggested that the $A_4$ peptide precursor may be related to a circulating immunoglobulin[44]. It has been demonstrated that CSF concentration of the high-molecular weight immunoreactive $A_4$ amyloid peptide precursor is about ten times higher than the normal concentration of plasma proteins[44], which can be indicative of the selective transport of this peptide across the blood–brain barrier.

The therapeutic effect of neuropeptides was also investigated in *dementia*, the most serious and common psychiatric disorder of old age, found also in Alzheimer disease and in senile dementia of the Alzheimer type[45]. Human studies with VP(Arg), VP(Lys), and their synthetic analogues DDAVP and DGAVP, have indicated that vasopressin and its analogues may improve memory and learning in apparently non-impaired adults[46]. Similar findings have been reported in patients with psychiatric and neurological disorders, such as *amnesia* of both alcoholic and post-traumatic origin, after electroconvulsive therapy, *Korsakoff* syndrome, Lesch–Nyhan disease, depression and memory disturbances concurrent with diabetes insipidus[47]. However, clinical trials of the VP-peptide family in Alzheimer disease and senile dementia of Alzheimer type gave results, which were negative, and at the best questionable[48]. Many of these studies were carried out on patients with complex neuropsychological deficits, other symptoms or profound brain degeneration, so it is possible that this degeneration has destroyed the recognition sites for action of VP-peptides. It is these complications which may be the cause of failure of this family of peptides to act in patients with extensive neuronal degeneration[49]. On the other hand, the positive therapeutic effects of VP-peptides in unimpaired subjects, as well as in different types of amnesia, strongly suggest that VP-like peptides do have an effect on the cognitive functioning of the brain, although the mechanism of action on specific altered processes remains to be further studied[46]. Thus it seems that these negative findings with VP-peptides in the treatment of dementia were inadequate in various aspects and it may be that the studies which had a positive result will be found to have a therapeutic relevance in the future[48].

The effects of ACTH-related peptides have also been studied in dementia[50–52]. The psychotropic response properties of the ACTH fragment, $ACTH_{4-10}$, have been examined, as well as that of the orally active synthetic analogues $ACTH_{4-9}$ sequence, and Org 2766, which is free of any peripheral endocrine side-effects of ACTH. In clinical trials $ACTH_{4-10}$ and Org 2766 have been studied in connection with a range of 'cognitive performance' tasks, and various psychophysiological indices were estimated. These investigations failed to demonstrate any consistent effect of these peptides on the performance of subjects in memory tests and some other tests for cognitive performance, apart from those requiring sustained vigilance[52,53]. Org 2766 shows a consistent positive therapeutic effect on mood that leads to increased sociability, feeling of competence and of greater cooperation in institutional life[48]. When Org 2766 was tested in patients with senile type of Alzheimer disease, some positive effects were obtained on cognitive and somatic functions[50]. However, several design problems in the clinical trials published to date, prevent any conclusions with regard to the therapeutic efficacy of VP-like and ACTH-like peptides for the treatment of dementia in Alzheimer disease and its senile type. Most of the problems have arisen from the difficulties of defining a homogenous diagnostic group with the same degree of dementia in all patients[48]. Decreased as well an increased VP(Arg) levels have been found in CSF in patients with Alzheimer disease[54–56]. On the other hand, there is a marked decrease in the volume of the VP(Arg) suprachiasmatic nuclei and in the total cell number in the nucleus in 80–100-year-old patients postmortem, and these changes were even more pronounced in patients with Alzheimer disease[57]. The observed decrease in the volume of suprachiasmatic nuclei in senescence and Alzheimer disease has been suggested to be a causal relationship in disturbances of the circadian rhythmicity of the sleep–wake cycle[58]. Similar analysis in trans-sexuals and Prader–Willi syndrome showed an extremely high number of VP(Arg) cells in the suprachiasmatic nucleus as well as an enlarged nucleus[58]. It has been suggested that a lack of sex hormones in certain stages of brain development may induce a large suprachiasmatic nucleus and observed increase in VP(Arg) cell bodies.

High concentrations of several neuropeptides are to be found in the basal ganglia of the human brain, including the *corpus striatum*, i.e. the caudate nucleus and the lentiform nucleus (the putamen and globus pallidus), the *claustrum*, and the area called '*ventral striatum*', i.e. nucleus accumbens, olfactory tubercle and the ventral part of the striatum and '*ventral pallidum*', i.e. ventral extension of globus pallidus with the rostral and subcommisural part of substantia innominata[59]. In these regions peptides are found which include the enkephalins, substance P, CCK, neurotensin, TRH, SRIF, VIP, angiotensin II, endorphins, etc. (Table 8.2). It is reasonable to assume that the peptides present in the structures of the extrapyramidal system may play a part in regulation of human motor behaviour. Furthermore, neuropathological studies have shown a significant alteration in neuropeptide concentrations in the basal ganglia, or within the afferent and efferent basal ganglia pathways in the brains of patients suffering from extrapyramidal disorders[59]. The majority of evidence has been provided

**Table 8.2** Concentration of neuropeptides in the basal ganglia and other regions of the human brain (according to Takagi, 1986[59])

| | *Brain region* | | | | | | | |
|---|---|---|---|---|---|---|---|---|
| | | *Globus pallidus* | | *Sub-stantia* | | | | |
| *Peptide* | *Caudate nucleus* | *Lateral* | *Medial* | *nigra* | *Amyg-dala* | *Hypo-thalamus* | *Cere-bellum* | *Cerebral cortex* |
| Enk(Met) (pmol/g) | 116 | 1163 | 675 | 661 | 26 | 141 | – | 42 |
| SP (pmol/g) | 138 | 197 | 877 | 1535 | 25 | 112 | – | 14 |
| CCK (pmol/g)* | 167 | – | – | – | 280 | 74 | 1 | 214 |
| VIP (pmol/g) | 5 | 2 | | 2 | 21 | 23 | 0 | 8 |
| SRIF (ng/mg)** | 0.6 | – | | 0.3 | 3.7 | 73.1 | 0.4 | – |
| TRH (ng/g)** | 21 | – | | – | 23 | 489 | 2 | – |
| ACE (pmol/min per mg) | 307 | 155 | 149 | 105 | – | – | 13 | 18 |
| NT (pmol/g) | 3 | 10 | | – | 5 | 43 | 1 | 1 |

*,** denotes values obtained in the guinea-pig* and rat**. ACE, angiotensin converting enzyme

from postmortem studies in Parkinson and Huntington disease and levels of enkephalin-like immunoreactivity are significantly decreased in the substantia nigra of patients with Huntington disease[60,61] or with striato-pallidal infarction[62]. On the other hand although the role of SRIF in the basal ganglia has not yet been fully understood, RIA studies have detected an increased level of this peptide in the striatum and globus pallidus of patients with Huntington disease[63,64]. It has also been suggested that SRIF may be involved in the local regulation of striatal dopaminergic transmission[65,66], while animal experiments strongly suggest that peripheral administration of SRIF can change the locomotor activity[67]. In addition, it has been found that the level of neurotensin in the striatum of patients with Huntington disease is also increased[68,69]. An increase of TRH-like immunoreactivity has also been detected in the striatum in Huntington disease, which may indicate the involvement of TRH in the dysfunction of the basal ganglia in this disease[69,70]. A marked decrease in angiotensin converting enzyme-like immunoreactivity has also been found in the striatum and substantia nigra in Huntington disease[71,72]. Significant reductions of enkephalins, substance P, VIP and CCK-8 have been reported for various parts of the striato-nigral and striato-pallidal systems in patients with this disease[73,74].

It has been shown that both Parkinson and Huntington disease are characterized by major alterations in 'classical' neurotransmitter concentrations in the basal ganglia, and it has been proposed that some form of interaction between neuropeptides and 'classical' neurotransmitters may play a role in pathophysiology of both disorders. This hypothesis has been supported by double staining or 'mirror' techniques at the electronmicroscopic level, which are useful to establish the interactions of the neuronal structures that contain different peptides, or to show the co-existence of multiple putative transmitters in single cells. Of particular interest, for extrapyramidal pathophysiology, is the coexistence of dopaminergic–peptidergic interactions and, in particular, between human mesencephalic dopaminergic neurons with enkephalinergic terminals[75], and between dopamine and CCK-8 in the cells of dopaminergic mesolimbic pathway, which has been shown to be affected in Parkinson disease[74]. The clinical importance of dopaminergic–peptidergic interactions in extrapyramidal disorders lies mainly in the possibility of a neuropharmacological approach, since orally active analogues of the peptides, with sufficient half-lives and blood–brain barrier penetration, may be employed to modify the altered dopaminergic activity in Parkinson disease and Huntington chorea[76]. Some therapeutic attempts have already been made by studying the effects of an orally active synthetic analogue Enk(Met),D-Ala$^2$-MePhe$^4$-Met-(O)-olo-enkephalin (FK-33-824) on the choreic movements. A significant reduction in the frequency of involuntary movements has been observed in three out of five patients between 60 and 90 min after drug administration[76]. In the same study, four patients with tardive dyskinesia were also treated with FK 33-824, but no changes in the symptoms of their disorders was observed. In other studies with double-blind cross-over controls using a single dose of FK 33-824, no positive therapeutic effects could be obtained in choreic patients or in Parkinson disease, generalized dystonia, torticollis, or tardive dyskineasia[76,77]. The different results between these studies may be due to administration of different doses of FK 33-824 needed to produce effects on the extrapyramidal system. The results indicate an enkephalinergic inhibition of striatal dopaminergic hyperactivity in Huntington disease, although the exact site of such inhibition has to remain speculative. In an earlier study the deteriorating effect of MSH injections on the symptoms of Parkinson disease has been demonstrated, and in a subsequent study with MSH-release inhibiting factor (MIF) a clinical global improvement in nine out of ten patients was shown[76]. The clinical improvement was seen on the third infusion day with MIF, and a better response was obtained with akinetorigid symptoms than with the tremor. The results obtained with MIF in Parkinson disease, by several groups, indicate central dopaminergic potentiation by this tripeptide, probably by modifying the striatal dopamine receptor sensitivity. The above studies suggest some therapeutic advantage of these peptides with regard to the systemic administration in patients with extrapyramidal symptoms, which may encourage further development in the clinical pharmacology of extrapyramidal disease[76].

The present data on the CSF concentrations of neuropeptides in extrapyramidal disorders, although limited, may contribute to our knowledge of the interactions between peptidergic and classical neurotransmitters in these disorders, as well as having some diagnostic or prognostic values. For example Enk(Met) concentration in the CSF of patients with Parkinson disease was found to be either normal or elevated in comparison to the control group, which was dependent on the phase of the disease and particularly with respect to the duration[77]. On the other hand, rather controversial results have been claimed about the concentration of substance P in the CSF of patients with Parkinson disease: one study showed no alteration while the other reported an elevation of this peptide in the CSF[78,79]. The latter finding was explained in terms of compensatory hyperactivity of an excitatory striato-nigral substance P pathway in response to the nigral dopaminergic hypoactivity of parkinsonians. No change in CSF concentration of substance P was observed in Huntington chorea[79]. From these preliminary findings it would appear that the CSF concentration of some 'classical' neurotransmitters or their metabolites may reflect the corresponding changes within the brain in some extrapyramidal diseases; this area of clinical research is very promising and may lead to a better understanding of the role of neuropeptides in extrapyramidal disorders.

To summarize, we may emphasize that at present there is sufficient clinical evidence to conclude that neuropeptides play an important role in normal and pathological human psychomotor behaviour. The function of the blood–brain and blood–CSF barrier and the interaction with peptides in various psychiatric and neurological disorders is probably at best understood with regard to the therapeutic effects of some stable peptide analogues and antagonists. However, the transport, metabolic and signal transduction functions of the blood–brain barrier and its relation to peptides is much less understood in brain disease from both the clinical and experimental point of view. The role of peptides in the regulation of blood–brain barrier permeability to various substrates is beginning to show that these molecules are important, both in the normal functioning of the human brain and in psychiatric and neurological disorders.

## REFERENCES

1. Prange, A.J., Gazzbutt, J., Loosen, P.T., Bissette, G. and Nemeroff (1987). The role of peptides in affective disorders: a review. *Prog. Brain Res.*, **72**, 235–279
2. Nemeroff, C.B., Widerlov, E., Bissett, G. Walleus, H., Karlsson I, Klund, K., Kilts, C.D., Loosen, P.T. and Vale, W.W. (1984). Elevated concentrations of CSF corticotropin-releasing factor-like immuno-reactivity in depressed patients. *Science*, **226**, 1342–1344
3. Kirkegaard, C.F., Hummer, L. and Rogowski, P. (1979). Increased levels of TRH in cerebrospinal fluid from patients with endogenous depression. *Psychoneuroendocrinology*, **4**, 227–235
4. Rimon, R., Le Greves, P., Nyberg, F., Heikkila, L., Salmela, L. and Terenius, L. (1984). Elevation of substance P-like peptides in the CSF of psychiatric patients. *Biol. Psychiatr.*, **19**, 509–516
5. Nemeroff, C.B. and Bissett, G. (1986). Neuropeptides in psychiatric disorders. In: *The American Handbook of Psychiatry*, Vol VIII, eds. Berger, P.A. and Brodie, H.K.H., Basic Books Inc., New York, 64–110

6. Gerner, R.H. and Yamada, T. (1982). Altered neuropeptide concentrations in cerebrospinal fluid of psychiatric patients. *Brain Res.*, **238**, 298–302
7. Rubinow, D.R., Gold, P.W., Post, R.M., Ballenger, I.C., Cowdry, R., Ballenger, I. and Reichlin, S. (1983). CSF somatostatin in affective illness. *Arch. Gen. Psychiat.*, **40**, 409–412
8. Bissette, G., Walleus, A., Widerlov, E., Karisson, I., Eklund, K., Loosen, P.T. and Nemeroff, C.B. (1984). Reductions of cerebrospinal fluid concentrations of somatostatin-like immunoreactive in dementia, major depression and schizophrenics. *Soc. Neurosci. (Abst)*, **10**, 1093
9. Gold, P.W., Ballenger, J.C., Robertson, G.L., Weingartner, H., Rubinow, D.R., Hoban, M.C., Goodwin, F.K. and Post, R.M. (1984). Vasopressin in affective illness: Direct measurement, clinical trials and response to hypertonic saline. In: *Neurobiology of Mood Disorders*, eds. Post, R.M. and Ballenger, J.C., Williams and Wilkins, Baltimore, 323–339
10. Verbanck, P.M.P., Lotstra, F., Gilles, C., Linkowski, P., Mendlewicz, J. and Vanderhaeghen, J.J. (1984). Reduced cholecystokinin immunoreactivity in the cerebrospinal fluid of patients with psychiatric disorders. *Life Sci*, **34**, 67–72
11. Gjerris, A.G., Fahrenkrug, J., Bojholm, S. and Rafaelsen, O.J. (1981). Vasoactive intestinal polypeptide (VIP) in cerebrospinal fluid in psychiatric disorders. In: *Biological Psychiatry*, eds. Perris, C., Struwe, G. and Jansson, B., Elsevier/North-Holland Biomedical Press, Amsterdam, 359–362
12. Carman, J.S., Post, R.M., Ballenger, J.C. and Goodwin, F.K. (1981). Reduced CSF calcitonin in mania. In: *Biological Psychiatry*, eds. Perris, C., Struwe, G. and Jansson, B., Elsevier/North-Holland Biomedical Press, Amsterdam, 457–461
13. Prange Jr, A.J., and Wilson, I.C. (1972). Thyrotropin releasing hormone (TRH) for the immediate relief of depression., A. preliminary report. *Psychopharmacologia*, **26**, 82
14. Prange Jr, A.J., Wilson, I.C., Lara, P.P., Alltop, L.B. and Breese, G.R. (1972). Effects of thyrotropin releasing hormone in depression. *Lancet*, **ii**, 999–1002
15. Kastin, A.J., Ehrensing, R.H., Schalch, D.S. and Anderson, M.S. (1972). Improvement in mental depression with decreased thyrotropin response after administration of thyrotropin releasing hormone. *Lancet*, **ii**, 740–742
16. Loosen, P.T. and Prange Jr, A.J. (1982). Serum thyrotropin response to thyrotropin-releasing hormone in psychiatric patients: a review. *Am. J. Psychiatr.*, **139**, 405–416
17. Loosen, P.T. (1985). The TRH-induced TSH response in psychiatric patients: a possible neuroendocrine marker. *Psychoneuroendocrinology*, , **10**, 237–260
18. Garbutt, J.C., Loosen, P.T., Blachrash, J. and Prange Jr, A.J. (1986). The prolactin response to thyrotropin releasing hormone in depressed patients and normal subjects. *Psychoneuroendocrinology*, **11**, 213–219
19. Maeda, K. (1976). Effects of thyrotropin-releasing hormone on growth hormone release in normal subjects and in patients with depression, anorexia nervosa and acromegaly. *Kobe J. Med. Sci.*, **22**, 263–272
20. Hsiao, J.K., Garbutt, J.C., Loosen, P.T., Mason, G.A. and Prange Jr, A.J. (1986). Is there paradoxical growth hormone response to thyrotropin-releasing hormone in depression? *Biol. Psychiatr.*, **21**, 595
21. Gold, P.W. and Chrousos, G.P. (1985). Clinical studies with corticotropin releasing factor: implications for diagnosis and pathophysiology of depression, Cushing's disease, and adrenal insufficiency. *Psychoneuroendocrinology*, **10**, 401–419
22. Amsterdam, J.D., Winokur, A., Abelman, E., Lucki, I. and Ricketls K (1983). Cosyntropin ($ACTH_{1\text{-}24}$) stimulation test in depressed patients and healthy subjects. *Am. J. Psychiatr.*, **140**, 907–908
23. Winokur, A., Amsterdam, J., Caroff, S., Snyder, P.J. and Brunswich, D. (1982). Variability of hormonal responses to a series of neuroendocrine challenges in depressed patients. *Am. J. Psychiatr.*, **139**, 39–44
24. Krahn, D., Meller, W.H., Shafer, R.B. and Morley, J.E. (1985). Cortisol response to vasopressin in depression. *Biol. Psychiatr.*, **20**, 918–921

25. Carrol, B.J., Davis, B. and Mowbray, R.M. (1972). *Depressive Illness: Some Research Studies*, Thomas, C.C., Springfield IL
26. Huey, L.Y., Janowsky, D.S., Mandell, A.J., Judd, L.L. and Pendery M (1975). Preliminary studies on the use of thyrotropin-releasing hormone in manic states, depression, and the dysphoria of alcohol withdrawal. *Psychopharmacol. Bull.*, **11**, 24–27
27. Yarbrough, G.G. (1976). TRH potentiates excitatory actions of acetylcholine in cerebral cortical neurons. *Nature (London)*, **263**, 523–524
28. Gerner, R.H., Catlin, D.H., Gorelick, D.A., Hui, K.K. and Li, C.H., (1980). β-endorphin: intravenous infusion causes behavioral change in psychiatric inpatients. *Arch. Gen. Psychiat.*, **37**, 642–647
29. Catlin, D.H., Gorcheck, D.A., Gerner, R.H., Hui, K.K. and Li, C.H. (1980). Clinical studies with human β-endorphin . In: *Polypeptide Hormones*, eds. Beers Jr, R.F. and Bassett, E.G., Raven Press, New York, 337–345
30. Pickar, D., Davis, G.C., Schulz, S.C., Extein, I., Wagner, R., Naber, D., Gold, P.W., Van Kammen, D.P., Goodwin, F.K., Wyatt, R.J., Li Ch H and Bunney, W.E. (1981). Behavioral and biological effects of acute injection in schizophrenic and depressed patients. *Am. J. Psychiatr.*, **138**, 160–166
31. Van der Velde, C.D. (1983). Rapid clinical effectiveness of MIF-I in the treatment of major depressive illness. *Peptides*, **4**, 297–300
32. Klein, H., Jungkunz, G., Nedopil, N., Ruther, E. and Spiegel, R. (1981). Clinical effects of the synthetic analogue of met-enkephalin FK-33-824. In: *Biological Psychiatry*, eds. Perris C, Struwe, G. and Jansson B. Elsevier/North-Holland Biomedical Press, Amsterdam, 390–393
33. Van Ree,, J.M. and De Wied,, D. (1981). Endorphins in schizophrenia. *Neuropharmacology*, **20**, 1271–1277
34. De Wied,, D. (1978). Psychopathology as a neuropeptide dysfunction. In: *Characteristics and Function of Opioids*, eds. Van Ree, J.M. and Terenius L. Elsevier, Amsterdam, 113–122
35. Van Ree, J.M., Verhoven, Z.K. and de Wied, D. (1987). Animal and clinical research on neuropeptides and schizophrenia. *Prog. Brain Res.*, **72**, 249–267
36. Van Ree, J.M., Caffe, A.M. and Wolterink, G. (1982). Non-opiate beta-endorphin fragments and dopamine. III. γ-Type endorphins and various neuroleptics counteract the hypoactivity elicited by injection of apomorphine into the nucleus accumbens. *Neuropharmacology.*, **21**, 1111–1117
37. Van Ree, J.M., Verhoeven, W.M.A. and De Wied, D. (1985). γ-Type endorphins: neurolepticum-like and anti-psychotic action. *Prog. Neuro-Psychopharmacol. Biol. Psychiatr.*, **9**, 561–567
38. Verhoeven, W.M.A., Van Ree, J.M., Verhey, F.H.M., Filip, V. and De Wied, D. (1985). The antipsychotic and neuroleptic-like action of γ-type endorphins. *Psychopharmacology: Impact on Clinical Psychiatry*, ed. Morgan, D. W. Ishiyaku EuroAmerica, St. Louis, Tokyo, 73–98
39. Itoh, H., Tanoue, S., Yagi, G., Tateyama, M., Kamisada, M., Fujii Y, Takamiya, M. and Nakajima, S. (1982). Clinical study on the psychotropic effects of caerulein an open clinical trial in chronic schizophrenic patients. *Keio J. Med.*, **31**, 71–95
40. Nair, N.P.V., Bloom, D.M. and Nestoros, J.N. (1982). Cholecystokinin appears to have anti-psychotic properties. *Prog. Neuropsychopharmacol. Biol. Psychiatry*, **6**, 509–512
41. Bloom, D.M., Nair, N.P.V. and Schwartz, G. (1983). CCK-8 in the treatment of chronic schizophrenia. *Psychopharmacol. Bull.*, **19**, 361–363
42. Mattes, J.A., Hom, W., Rochford, J.M. and Orlosky, M. (1985). Ceruletide for schizophrenia: a double-blind study. *Biol. Psychiatry*, **20**, 533–538
43. Van Ree, J.M., Verhoeven, W.M.A., Brouwer, G.J. and De Wied, D. (1984). Ceruletide resembles anti-psychotics in rats and schizophrenic patients. *Neuropsychobiology.*, **12**, 4–8

44. Pardridge, W.M., Vinters, H.V., Yang, J., Eisenberg, J., Choi, T.B., Tourtelotte, W.W., Huebuer, V. and Shirley, J.E. (1987). Amyloid angiopathy of Alzheimer's disease: amino acid compositione and partial sequence of a 4,200–dalton peptide isolated from cortical microvessels. *J. Neurochem*, **49**, 1394–1401
45. Huppert, F.A., Tym, E. (1986). Clinical and neuropsychologic assessment of dementia. *Br. Med. Bull.*, **42**, 11–18
46. Beckwith, B.E., Till, R.E. and Schneider, V. (1985). Vasopressin analogue (DDAVP) improves memory in human males. *Peptides*, **5**, 819–822
47. Tinklenberg, J.R. and Thornton, J.E. (1983). Neuropeptides in geriatric psychopharmacology. *Psychopharmacol. Bull.*, **19(2)**, 198–211
48. Kragh-Sorensen, P.A. and Lolka, A. (1987). Neuropeptides and dementia. *Prog. Brain Res.*, **72**, 269–277
49. De Wied, D. and Jolles, J. (1982). Neuropeptides derived from pro-opiocortin: behavioural, physiological and neurochemical effects. *Physiol. Rev.*, **62(3)**, 976–1059
50. Kragh-Sorensen, P., Bank Olsen, R., Lund, S., Van Riezen, H. and Steffensen, K. (1986). Neuropeptides: ACTH-peptides in dementia. *Neuro-psychopharmacol. Biol. Psychiatr.*, **10**, 479–492
51. Pigache, R.M. (1982)., A. peptide for the aged? Basic and clinical studies. In: *Psychopharmacology of Old Age*, ed. Wheatley D. Oxford University Press, Oxford, 67–96
52. Pigache, R.M. (1983). The human psychopharmacology of peptides related to ACTH and alpha-MSH. In: Gram, Usdin, Dahl, Kragh-Sørensen, Sjöqvist and Morselli (Eds.), *Clinical Pharmacology in Psychiatry*, Vol. 3. McMillan Press Ltd., London and Basingstoke, 361–377
53. Pigache, R.M. and Rigter, H. (1981). Effects of peptides related to ACTH on mood and vigilance in man. *Front Hormone Res.*, **8**, 193–207
54. Raskind, M.A., Peskind, E.R., Lampe, T.H., Risse, S.C., Taborsky, G.J. and Dorsa, D. (1986). Cerebrospinal fluid vasopressin, oxytocin, somatostatin and beta-endorphin in Alzheimer's disease. *Arch Gen. Psychiatry*, **43**, 382–388
55. Sørensen, P.S., Hammer, M., Vorstrup, S. and Gjerris, F. (1983). CSF and plasma vasopressin concentrations in dementia. *J. Neurol. Neurosurg. Psychiatry*, **46**, 911–916
56. Sundquist, J., Forsling, M.L., Olsson, J.A. and Akerlund, M. (1983). Cerebrospinal fluid arginine vasopressin in degenerative disorders and other neurological disease. *J. Neurol. Neurosurg. Psychiatry*, **46**, 14–17
57. Swaab, D.F., Fliers, E. and Partiman, T.S. (1985). The suprachiasmatic nucleus of the human brain in relation to sex, age and senile dementia. *Brain Res.*, **342**, 37–44
58. Swaab, D.F., Roozendaal, R.R., Velis, D.N., Goorn, L. and Williams, R.S. (1987). Suprachiasmatic nucleus in ageing, Alzheimer's disease, transexuality and Prader-Willi syndrome. *Prog. Brain Res.*, **72**, 301–310
59. Takagi, H. (1986). Distribution of peptides in basal ganglia. *Prog. Brain Res.*, **66**, 35–72
60. Emson, P.C., Arregui, A., Clement-Jones, V., Sandberg, B.E.B. and Rossor, M. (1980a). Regional distribution of methionine-enkephalin and substance P-like immunoreactivity in normal human brain and in Huntington's disease. *Brain Res.*, **199**, 147–160
61. Marshall, P.E. Landis, D.M. and Zalneraitis, E.L. (1983). Immunocytochemical studies of substance P and leucine-enkephalin in Huntington's disease. *Brain Res.*, **289**, 11–26
62. Pioro, E.P., Hughes, J.T. and Cuello, A.C. (1984). Loss of substance P and enkephalin immunoreactivity in the human substantia nigra after striato-pallidal infarction. *Brain Res.*, **292**, 339–347
63. Aronin, N., Cooper, P.E., Lorenz, L.J., Bird, E.D., Sagar, S.M., Leeman, S.E. and Martin, J.B. (1983). Somatostatin is increased in the basal ganglia in Huntington's disease. *Ann. Neurol.*, **13**, 519–526
64. Nemeroff, C.B., Youngblood, W.M., Manberg, P.J., Prange, A.J. and Kizer, J.S. (1983b). Regional brain concentrations of neuropeptides in Huntington's chorea and schizophrenia. *Science*, **221**, 972–975

65. Chesselet, M.F. and Reisine, D. (1983). Somatostatin regulates dopamine release in rat striatal slices and cat caudate nuclei. *J. Neurosci.*, **3**, 232–236
66. Beal, M.F. and Martin, J.B. (1984). The effect of somatostatin on striatal catecholamines. *Neurosci. Lett*, **44**, 271–276
67. Cohn, M.L. and Cohn, M. (1975). "Barrel rotation" induced by somatostatin in the non-lesioned rat. *Brain Res.*, **96**, 138–141
68. Manberg, P.J., Nemeroff, C.B., Iversen, L.L., Rossor, M.N., Kizer J, S. and Prange, A.J. (1982). Human brain distribution of neurotensin in normals, schizophrenics, and Huntington's choreics. *Ann. N.Y. Acad. Sci.*, **400**, 354–367
69. Nemeroff, C.B., Luttinger, D., Hernandez, D.E., Mailman, R.B., Mason, G.A., Davis, S.D., Wilderlov, E., Frye, G.D., Kilts, C.A., Beaumont K, Breese, G.R. and Prange, A.J. (1983). Interactions of neurotensin with brain dopamine systems: biochemical and behavioral studies. *J. Pharmacol. Exp. Ther.*, **225**, 337–345
70. Spindel, E.R., Wurtman, R.J. and Bird, E.D. (1980). Increased TRH content of the basal ganglia in Huntington's disease. *N. Engl. J. Med.*, **303**, 1235–1236
71. Arregui, A., Bennett, J.P., Bird, E.D., Yamamura, H.I., Iversen, L. L and Snyder, S.H. (1977). Huntington's chorea: selective depletion of activity of angiotensin converting enzyme in the corpus striatum. *Ann. Neurol.*, **2**, 294–298
72. Arregui, A., Emson, P.C. and Spokes, E.G. (1978). Angiotensin-converting enzyme in substantia nigra: reduction of activity in Huntington's disease and after intrastriatal kainic acid in rats. *Eur. J. Pharmacol.*, **52**, 212–124
73. Agid, Y. (1985). Biochemical neuropathology: A new approach to Parkinson's disease. In: *Clinic Experiences with Budipine in Parkinson Therapy*, eds. Gerstenbrand, F., Poewe, W. and Stern G. Springer, Heidelberg, 4–12
74. Agid, Y. and Javoy-Agid, F. (1985). Peptides and Parkinson's disease. *Trends Neurosci.*, **8**, 30–35
75. Gaspar, P., Berger, B., Gay, M., Hamon, M., Cesselin, F., Vigny, A., Javoy-Agid, F. and Agid, Y. (1983). Tyrosine hydroxylase and methionine-enkephalin in the human mesencephalon. *J. Neurol. Sci.*, **58**, 247–267
76. Gerstenbrand, F. and Poewe, W. (1986). Dopaminergic-peptidergic interactions in extra pyramidal disorders: a review of clinical evidence. *Adv. Neurol.*, **45**, 67–73
77. Sheehy, M., Schachter, M., Marsden, C.D. and Parkes, J.D. (1981). Enkephalins in motor disorders. In: *Research Progress in Parkinson's Disease*, eds. Rose, F.C. and Capildeo R. Pitman Medical, London, 165–169
78. Pezzoli, G., Panerai, A.E., Digiulio, A., Longo, A., Passerini, D. and Carenzi, A. (1984). Methionine-enkephalin substance P and acid in the CSF of Parkinsonian patients. *Neurology*, **34**, 516–519
79. Nutt, J.G., Mroz, E.A., Leeman, S.E., Williams, A.C., King Engle W and Chase, T. (1980). Substance P in human cerebrospinal fluid: reductions in peripheral neuropathy and autonomic dysfunction. *Neurology*, **30**, 1289–1293

# 9 Immunological and other aspects of the blood–brain barrier

## BRAIN AND IMMUNE SYSTEM

It has been demonstrated that the central nervous system and the immune system have several characteristics in common, and that they are functionally interconnected in both ways, namely the brain may influence the immune system, and the activation of the immune system by antigenic stimulation may induce changes in brain function and in neuroendocrine profiles[1]. One of the classical examples with respect to the influence of the psychosocial factors on the immunocompetence in humans is the effect of bereavement on immune function. It has been shown that bereavement induced suppression of mitogen-induced T and B cell responses in cell cultures compared to pre-bereavement levels[2–4]. Furthermore, it has been shown that an important element in psychosocial stress is closely related to the ability of the individual to cope with the situation, which is reflected in either significant reduction of mitogen-induced T cell activity, or in significantly lower natural killer (NK) cell activity in individuals, who do not cope with psychic distress[5,6]. Similar findings were observed in animal experiments in which the animals were exposed to stress and changes in immune responsiveness were found[7]. It has been accepted that two main pathways are involved in the mechanisms of the neuro-immunomodulation[1]. The first pathway is *humoral* and involves the hypothalamic–pituitary–adrenal axis and corresponding peptides such as enkephalins, endorphins, ACTH, VIP, as well as prolactin, corticosteroids and sex hormones. The second pathway is *neural*, including the autonomic nervous system, innervation of the lymphoid tissues such as thymus, spleen, lymph nodes and even bone marrow[8,9]. Neurotransmitters and neuropeptides may locally modulate immune reactivity in this circuit, since the corresponding receptor sites have been identified on lymphocytes, and the presence of neurotransmitters–neuropeptides interaction has been reported to modulate the lymphocyte function[10,11]. Some studies have been directed

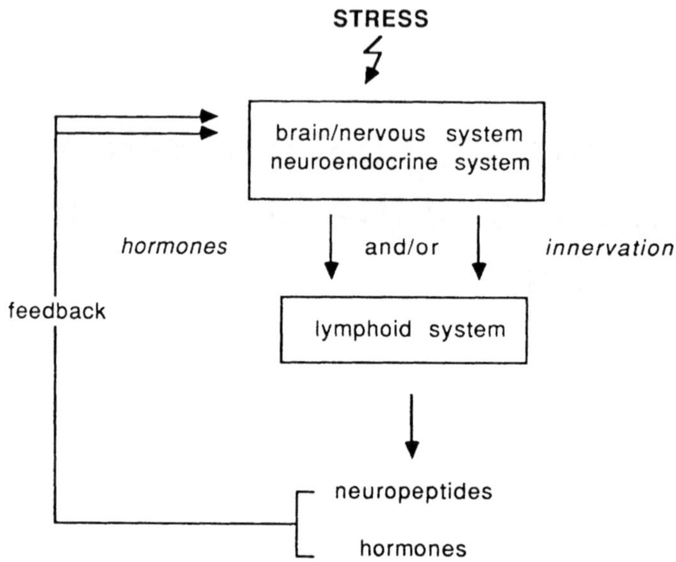


**Figure 9.1** Negative feedback regulation between brain and immune system mediated by neuropeptides and peptide hormones according to Ballieux and Heijnen (1987)[1]

directed towards immunomodulation of human lymphocyte by opioid peptides[12] and include clinical trials on the effects of enkephalins and endorphins in cancer and AIDS patients[12], as well as the effects of $\beta$-endorphin and several of its fragments on cell proliferation and antibody synthesis *in vitro*[13,14]. Although sometimes conflicting results were obtained, these studies generally confirmed the neuro-immunomodulatory role of opioid peptides.

It has been shown, on the other hand, that an ongoing immune response may generate signals which can be received by the brain, and result in an increased firing rate of neurons in some specialized brain areas, such as the ventromedial nucleus of the hypothalamus[15]. In subsequent studies, it was possible to demonstrate that factors, produced by lymphoid cells *in vitro* after stimulation with mitogen, have similar effects on hypothalamus, as were the ones previously obtained in the hypothalamus by an *in vivo* immune response[16]. These results strongly suggest that the lymphocyte products, called *lymphokines*, mediate the influence of the immune system on the central nervous system. Further investigations of these lymphoid cell derived 'neurohormone(s)' using specific antisera have demonstrated, on the basis of immunoreactivity, that human as well as rodent lymphocytes can produce a number of peptides, such as ACTH, growth factor, TSH, HCG and endorphin-like peptides[17]. Further mRNA for POMC in spleen lymphocytes and macrophages has been demonstrated[18,19], as well as mRNA for preproenke-

been shown that production of some lymphoid cell peptides undergoes similar regulation as in the hypothalamo–pituitary axis. For example, CRF may induce both *de novo* synthesis and release of leukocyte-derived $ACTH_{1-39}$ and $\beta$-endorphin, and in addition dexamethasone blocks the production of these peptides[21]. The presence of factors in lymphokines, which may influence the release of CRF in the hypothalamus, has been demonstrated as well as the subsequent rise in the circulating cortisol level[16,22]. Another factor isolated from macrophages, named interleukin 1, may induce the release of ACTH from the pituitary gland with a simultaneous increase in serum cortisol level[23].

Taking the available clinical and experimental data together it seems that neuropeptides and peptide hormones play a most important role as possible mediators between the nervous and immune systems. The possible feedback regulations may be presented schematically as in Figure 9.1. As can be seen, the brain and the neuroendocrine system influence the immune system by secreting neuropeptide and peptide hormone mediators, while the products of activated cells of the immune system are also neuropeptides and peptide hormones, which can feedback on the brain and endocrine system. It is therefore essential that before the actions of lymphokines on the brain can be fully understood, the complex interactions between these systems taking place at the blood–brain barrier must be carefully elucidated.

## IMMUNOGLOBULINS AND BLOOD–CSF BARRIER

The development and application of immunochemical and nephelometric methods and the introduction of the enzyme linked immunosorbent assays (ELISA), for the quantitative analysis of single proteins and immunoglobulins in the cerebrospinal fluid and serum, has led to a greater understanding of the immunological role of the blood–brain and blood–CSF barriers. These methods can be used to determine the altered levels of different classes of immunoglobulins in the CSF found in some neurological and psychiatric diseases, as well as in animal experimental models of these disorders[24]. By using these new methods, instead of electrophoretically determined globulin fractions, single immunoglobulins G, A and M fractions can be quantified, and, in particular, the ELISA method is sensitive and specific enough to identify antigen-specific antibodies and proteins in CSF with concentrations of less than 10 pg $ml^{-1}$. From the clinical viewpoint, evaluation of the blood–CSF barrier dysfunction to immunoglobulins in neurological diseases is based on an elevated level of immunoglobulin and/or their particular fractions in the CSF, expressed as the CSF/plasma quotient for these immunoglobulins, relative to the CSF/plasma quotient for albumin[25]. An estimate of the selectivity of the blood–brain and blood–CSF barriers to various proteins, as determined in the CSF, can now be expressed relative to albumin, and the quantification of the aberrant concentrations of immunoglobulins, as the *immunoglobulin* index. Under steady-state conditions, the molecular size dependent ratio of CSF/serum concentration quotients for different classes of immunoglobulins relative to albumin/CSF serum quotient has been determined in 150 patients with non-inflammatory neurological diseases, and this has been used as the control group for patients with neurological disorders with a possible immunological component[26]. The negative correlation between immunoglobulin molecular size and their CSF/plasma ratio, relative to albumin CSF/plasma quotient, agrees with the general correlation between the haemodynamic radius for proteins and the CSF/serum concentration ratio[26]. According to this hypothesis, the selectivity of the restraining barrier can be expressed by a selectivity coefficient, and any deviation of the immunoglobulin ratio from the barrier selectivity line can be taken to represent either blood–brain barrier dysfunction, with an increased permeability of the barrier to plasma proteins, or, alternatively, pathological local CNS immunoglobulin synthesis[24]. Thus, if an IgG quotient is increased without change in the albumin quotient this may indicate an *in situ* local synthesis of immunoglobulins in the brain. However, an increase of the albumin quotient followed by a corresponding increase in the IgG quotient may indicate a blood–CSF barrier dysfunction[24]. Combined disturbances would have an immunoglobulin G index in between these values[27]. Although this hypothesis is apparently straightforward for the clinical differentiation of neurological disorders with primary altered immunoglobulin synthesis in the CNS, or blood–CSF barrier dysfunction, one cannot rule out the possibility of an altered transport of immunoglobulins across the blood–brain barrier relative to albumin transport. This possibility must be considered with respect to recent findings of carrier and/or receptor mediated albumin transfer across the blood–brain barrier[28], and its regulation by the level of cyclic nucleotides in the cerebral capillaries[29]. Also, recent investigations of the immunoglobulin transport across the blood–brain barrier with $^{125}$I-labelled immunoglobulin G and vascular brain perfusion methods have revealed a time-dependent transfer[30], which has been confirmed by immunocytochemical methods. Taking this into account, it is possible to speculate that a specific increase of IgG quotient found in some neurological and psychiatric disorders may also represent a selective increase in transfer for that particular immunoglobulin across the blood–brain barrier.

## EXPERIMENTAL MODELS OF BRAIN DISORDERS

The neurobiological basis of psychiatric and neurological disorders has greatly expanded over the last two decades. A major part of this expansion, so far as the mechanisms at the cellular level are concerned, has been due to development of various animal experimental models of the diseases. Models for *non-brain disorders* often have a close similarity between the biological condition in man and that which can be created in the animal[31]. Many of these models are *homologous*, i.e. there is a correspondence between the etiology of the disease and the model of that disease. Other models may be only *isomorphic*, in that, although there is a parallelism between altered function in the model and the disease in the human, the etiology of the conditions may be quite different, as well as a direct cause of the altered conditions in humans and in the animal model. Finally, animal models may have no resemblance to the disease, i.e. it can be *non-homologous* and *non-isomorphic*, but may reflect some aspects of the disease and/or be valuable for studying the therapeutic effects of drugs. Experimental models in psychiatry are usually *non-homologous*, but in some of these models there is a *pharmacological isomorphism*, i.e. a drug induced behavioural syndrome in the animal which resembles the psychiatric conditions in man, or some aspect of the psychiatric disorder may be induced by application of drugs in the animal[32–35]. Experimental models in neurology are usually *isomorphic*.

The role of the blood–brain barrier related to peptide and protein in the central nervous system in various experimental models of brain disorders may be of considerable importance. For example, in the amphetamine model of *experimental 'psychosis'* a reversible functional alteration of the blood–brain barrier permeability has been reported[35], and a reversible increase of the unidirectional blood to brain transport of lipid-insoluble molecules such as mannitol and polyethylene glycol (MW, 4000) has been demonstrated. The kinetic features of this finding suggests that there has been an opening up of channels in the blood–brain barrier which permit a flow of fluid, which in turn may carry substances across the barrier at a rate irrespective of their molecular size. Furthermore, in the same model a marked increase of the initial volume of distribution for Enk(Leu) and VP(Arg) has been shown, while the cerebrovascular permeability for Enk(Leu) has not been altered in comparison to that found in normal non-treated animals, while that for VP(Arg) has increased by about two to three times[35]. On the other hand, cerebrovascular permeability to TRH seems to be unchanged, or only moderately increased, depending on the brain region, while the initial volume of distribution appears to be reduced[35]. These results may suggest the possibility of an induced carrier-mechanism for Enk(Leu) and VP(Arg), which needs to be further investigated.

Experimental evidence for close functional association between dopamine and neuropeptides in the extrapyramidal system has been obtained in animal studies of the neurotoxicity of N-methyl-4-phenyl-1,2,3,6-tetrahydropyridine (MPTP), – a synthetic pethidine derivative, causing the nigral cells to be damaged and producing a clinical syndrome which resembles Parkinsonism. In marmosets exposed to MPTP, a depletion of striatal dopamine and profound changes in the levels of substance P, VIP and neuropeptide Y were detected in the region of the substantia nigra[36].

Recent investigations of $^{125}$I-insulin binding to isolated brain microvessels from control, streptozotocin diabetic and insulin-treated diabetic rats indicate that a specific defect occurs in the diabetic brain capillary which may cause a further deficiency in brain insulin in poorly controlled diabetes[37]. Also, it has been shown that *in vitro* brain capillary endothelial cells have the capacity to increase their insulin receptors in response to a low-insulin environment, whereas *in vivo* the microvessels decrease their insulin receptors in response to diabetes.

A reversible increase in the blood–CSF barrier permeability to albumin has been demonstrated in both the guinea-pig and in the rat model of multiple sclerosis, following the acute phase of the chronic relapsing experimental allergic encephalomyelitis (EAE) induced by the homologous spinal cord homogenates[38,39]. On the other hand, neither in the acute phase nor during the relapsing phase of the disease was any local CNS synthesis of IgG demonstrated[38] so it seems that the blood–brain barrier permeability changes to IgG are not important. However, a note of caution must be sounded, since both in EAE and in multiple sclerosis in man, the disease is multifocal, and not diffuse, so the localized areas of the spinal cord are affected in different animals. However, the protein determinations were made from bulk CSF samples which represent the 'average' of the pathological processes from that particular area of CNS from which the CSF sample was drawn.

## MODULATING INFLUENCES ON PEPTIDE TRANSPORT ACROSS THE BLOOD–BRAIN BARRIER

Different modulating influences may act separately or additively on a given aspect of the blood–brain barrier permeability[40]. For example, *aluminium* has been shown to alter transmembrane diffusion across the blood–brain barrier without affecting its restrictive function or brain haemodynamics[41,42]. This subtle effect of aluminium has been suggested to underlay the etiopathogenesis of Alzheimer disease by increasing the blood–brain barrier permeability of DSIP. It has been recently demonstrated that aluminium as a neurotoxin is capable of altering blood–brain barrier structure and function with respect to the transfer of N-tyrosinated peptides, Tyr-MIF-1 and enkephalin from brain to blood[43]. This effect is due to non-competitive inhibition of this saturable peptide transport system, and for example the brain to blood transfer of albumin, D-Tyr-MIF-1 and technetium pertechnetate are not affected by aluminium. Findings from this study suggest that one possible mechanism by which aluminium may induce neurotoxicity is by selective alteration of the transport systems at the blood–brain barrier. *Lighting* conditions appear to be able to alter both transmembrane diffusion and saturable transfer at the blood–brain barrier. Specifically, constant darkness and reverse cycle lighting may increase the blood–brain barrier permeability to DSIP[44], and this has been demonstrated to be additive to changes induced by aluminium. On the other hand, constant lighting decreases the transport of Tyr-MIF-1 across the blood–brain barrier from CSF to blood, and this may be consistent with the rhythmic nature of these peptides with regard to both their levels and actions[45]. It seems that there is an inhibition of the brain to blood transport system for Tyr-MIF-1 and enkephalins in analgesia induced either by restraint or stress and/or by administration of morphine[46], but it has been suggested that these events probably represent two separate aspects of opiate action, that may even be mediated by separate receptor sites or peptides in the opiate family. Changes in blood–brain barrier permeability with ageing have also been reported for the transport system of small, N-tyrosinated peptides, from CSF to blood where both $K_m$ and $V_{max}$ were decreased by about 25% in 26 month old rats in comparison with 4-month-old rats[47]. Also an increased availability of circulating insulin to the newborn brain has been suggested, which is further corroborated by elevated CSF/serum and brain/serum insulin ratios in the new born vs. adult[37]. By means of the Oldendorf method, it has been shown that BUI for insulin is 22% for the newborn rabbit, while in 3-week-old and in adult rabbits, BUI values are 13% and 6.5%, respectively. Similarly, an increased specific insulin binding to isolated cerebral microvessels is found in the newborn rabbit, while a significant decrease of binding comparing to the newborn is found in 3-week-old and adult animals. These results suggest an increase in insulin receptor number in the newborn, with only minimal changes in affinity.

## REFERENCES


1. Ballieux, R.E. and Heijnen, C.J. (1987). Brain and immune system, a one-way conversation or a genuine dialogue? *Prog. Brain Res.*, **72**, 71–77
2. Bartrop, R.W., Luckhurst, E., Lazarus, L., Kiloh, L.B. and Penny R (1977). Depressed lymphocyte function after bereavement. *Lancet*, **i**, 834–937
3. Schleifer, S.J., Keller, S.E., Camerino, M., Thorton, J.C. and Stein, M. (1983). Suppression of lymphocyte stimulation following bereavement. *J. Am. Med. Assoc.*, **250**, 374–377
4. Stein, M., Schleifer, S.J. and Keller, S.E. (1986). Stress, depression and immunity. In: *Neuro-Regulation of Autonomic, Endocrine and Immune Systems*, eds, Frederickson, R.C.A., Hendrie, H. C, Hingtgen, J.N. and Aprison, M. H. Martinus Nijhoff, Boston, 367–388
5. Samuel, D., Naor, S., Pecht, M. and Trainin, N. (1986). Bereavement and the Immune system: the Effect of the Loss of an Unborn Child on Mitogenic Stimulation of, T. cells. Abstract No 64, *International Workshop on Neuroimmunomodulation*, Dubrovnik, June 1–6
6. Locke, S.E., Kraus, L., Leserman, J., Hurst, M.W., Heisel, J.S. and Williams, R.M. (1984). Life change stress, psychiatric symptoms, and natural killer cell activity. *Psychosom. Med.*, **46**, 441–453
7. Veldhuis, H.D., Croiset, G., Ballieux, R.E., Heijnen, C.J. and De Wied, D. (1986). *Proceedings of the Conference on Hypothalamic Dysfunction in Neuropsychiatric Disorders*, eds. Nerozzi, D. and Frajese G. Raven Press, New York
8. Felten, D.L., Felten, S.Y., Carlson, S.L., Olschowka, J.A. and Livnat, S. (1985). Noradrenergic and peptidergic innervation of lymphoid tissue. *J. Immunol.*, **135**, 755s–765s
9. Bulloch, K. and Pomerantz, W. (1984). Autonomic nervous system innervation of thymic-related lymphoid tissue in wild-type and nude mice. *J. Comp. Neurol.*, **228**, 57–63

10. Tecoma, E.S. and Huey, L.Y. (1985). Psychic distress and the immune response. *Life Sci.*, **36**, 1799–1812
11. Guillemin, R., Cohn, M. and Melnechuk, T. (eds.)(1985). *Neural Modulation of Immunity*. Raven Press, New York
12. Plotnikof, N.P., Miller, G.C., Nimeh, N., Faith, R.E., Murgo, A.J. and Wybran, J. (1987). Enkephalins and T-cell enhancement in normal volunteers and cancer patients. *Ann. N.Y. Acad. Sci.*, **496**, 608–619
13. Heijnen, C.J., Bevers, C., Kavelaars, A. and Ballieux, R.E. (1986). Effect of alpha-endorphin on the antigen-induced primary antibody response of human blood B cells *in vitro*. *J. Immunol.*, **136**, 213–216
14. Heijnen, C.J., Croiset, G., Zijlstra, J. and Ballieux, R.E. (1987). Modulation of lymphocyte function by endorphins. *Ann. N.Y. Acad. Sci*, **496**, 161–165
15. Besedovski, H.O., Sorkin, E., Felix, D. and Haas, H. (1977). Hypothalamic changes during the immune response. *Eur. J. Immunol.*, **7**, 323–325
16. Besedovski, H.O., Del Rey, A., Sorkin, E., Lotz, W. and Schwulera, U. (1985). Lymphoid cells produce an immuno-regulatory glucocorticoid increasing factor (GIF) acting through the pituitary gland. *Clin. Exp. Immunol.*, **59**, 622–628
17. Blalock, J.E. (1985). Proopiomelanocortin-derived peptides in the immune system. *Clin. Endocrinol.*, **22**, 823–827
18. Lolait, S.J., Clements, J.A., Markwick, A.J., Cheng, C., McNally M, Smith, I.A. and Funder, J.W. (1986). Pro-opiomelanocortin messenger ribonucleic acid and post-translational processing of beta endorphin in spleen macrophages. *J. Clin. Invest.*, **77**, 1776–1779
19. Westly, H.J., Kleiss, A.J., Kelley, K.W., Wong, P.K.Y. and Yuen, P. -H, (1986). Newcastle disease virus-infected splenocytes express the pro-opiomelanocortin gene. *J. Exp. Med.*, **163**, 1589–1594
20. Zurawski, G., Benedik, M., Kamb, B.J., Abrams, J.S., Zurawski, S.M. and Lee, F.D. (1986). Activation of mouse T helper cells induces abundant preproenkephalin mRNA synthesis. *Science*, **232**, 772–775
21. Smith, E.M., Morrill, E.M., Meyer III, W.J. and Blalock, J.E. (1986). Corticotropin releasing factor induction of leukocyte-derived immunoreactive ACTH and endorphins. *Nature (London)*, **321**, 881–882
22. Blalock, J.E., Bost, K.L. and Smith, E.M. (1985). Neuroendocrine peptide hormones and their receptors in the immune system. Production, processing and action. *J. Neuroimmunol*, **10**, 31–40
23. Del Rey, A. (1986). Immune-feedback regulatory signals. In: *Hormones and Immunity*, ed. Berczi, I. Elsevier, New York
24. Reiber, H. (1986). Evaluation of blood cerebrospinal fluid barrier dysfunctions in neurological diseases. In: *The Blood–Brain Barrier in Health and Disease*, eds. Suckling, A.J., Rumsby, M.G. and Bradbury, M.W. B. Ellis Horwood Ltd., Chichester, 132–147
25. Tibbling, C., Link, H. and Ohman, S. (1977). Principle of albumin and IgG analyses in neurological disorders. I. Establishment of reference values. *Scand. J. Clin. Lab. Invest.*, **37**, 385–390
26. Felgenhauer, K. (1974). Protein size and cerebrospinal fluid. *Klin. Wochen.*, **52**, 1158–1164
27. Felgenhauer, K. and Schädlich, H.J. (1987). The compartmental IgM and IgA responsee within the central nervous system. *J. Neurol. Sci.*, **77**, 125–135
28. Kumagai, A.K., Eisenberg, J. and Pardridge, W.M. (1987). Absorptive-mediated endocytosis of cationized albumin and a beta-endorphin-cationized albumin chimeric peptides by isolated brain capillaries. Model system of blood–brain barrier transport. *J. Biol. Chem.*, **262**, 15214–15219
29. Joó, F. (1988). Cyclic nucleotide-mediated regulation of albumin transport in brain microvessels. In: *Peptide and Amino Acid Transport Mechanisms in the Central Nervous*

*System*, eds. Rakic, Lj., Begley, D.J., Davson, H. and Zlokovic, B. V. Macmillan, London, 121–131

30. Zlokovic, B.V., Skundric, D.S., Segal, M.B., Colver, J., Jankov, R. M, Pejnovic, N., Lackovic, V., Mackic, J., Lipovac, M.V., Davson, H., Kasp E, Dumonde, D., Rakic, Lj. (1989). Blood–brain permeability changes during acute allergic encephalomytis induced in the guinea pig. *Metab. Brain Dis.*, **4**, 33–40
31. Kornetsky, C. (1977). Animal models: promises and problems. In: *Animal Models in Psychiatry and Neurology*, eds. Hanin, I. and Usdin E. Pergamon Press, Oxford, pp.1–9
32. Rakic, Lj., Zlokovic, B.V., Segal, M.B., Lipovac, M.H., Mitrovic D, M., Veskov, R., Mackic, J.B. and Davson, H. (1989). Effects of sensory-motor cortical lesions in blood–brain barrier permeability in guinea-pigs. *Metab. Brain Dis.*, **4**, 9–15
33. Prange, A.J., Gazzbutt, J., Loosen, P.T., Bissette, G. and Nemeroff, C.B. (1967). The role of peptides in affective disorders: a review. *Prog. Brain Res.*, **72**, 235–279
34. Van Ree, J.M. and Verhoeven, de Weid, D. (1987). Animal and clinical research on neuropeptides and schizophrenia. *Prog. Brain Res.*, **72**, 249–267
35. Rakic, Lj., Zlokovic, B.V., Davson, H., Begley, D.J., Segal, M.B., Lipovac, N., Mitrovic, D.M., Mackic, J.B. and Veskov, R. (1988). Experimental psychosis and transport of amino acid and peptides across the blood–brain barrier. In: *Peptide and Amino Acid Transport in the Central Nervous System*, eds. Rakic, Lj., Begley, D. J, Davson, H. and Zlokovic, B.V. Macmillan, London, 173–187
36. Allen, J.M., Cross, A.J., Yeates, J.C. Close, S., Pay, S., Marriott A, S., Tyers, M.J., Crow, T.J. and Bloom, S.R. (1985). Effect of MPTP on adrenal and central neuropeptides and monoamines in the marmoset. *J. Endocrinol.*, **104**, 196–199
37. Frank, H.J.L., Jankovic-Vokis, T., Pardridge, W.M. and Morris, W.L. (1985). Enhanced insulin binding to blood–brain barrier *in vivo* and to brain microvessels *in vitro* in newborn rabbits. *Diabetes*, **34**, 728–733
38. Suckling, H.J. and Rumsby, M.C. (1986). The blood–cerbrospinal fluid barrier in chronic relapsing experimental allergic encephalomyelitis. In: *The Blood–Brain Barrier in Health and Disease*, eds. Suckling, A.J., Rumsby, M.E. and Bradbury, M.W.B., Ellis Harwood, Chichester, 132–147
39. Daniels, P.M., Lam, D.K. and Pratt, O.E. (1983). Relation between the increase in diffusional permeability of the blood–central nervous system barrier and other changes during the development of experimental allergic encephalitis in the Lewis rat. *J. Neurol. Sci.*, **60**, 367–376
40. Banks, W.A. and Kastin, A.J. (1988). Peptides and the blood–brain barrier. In: *Peptides and Amino Acids in the Central Nervous System*, eds. Rakic, Lj., Begley, D.J., Davson, H. and Zlokovic, B.V. Macmillan, London, 21–32
41. Banks,, W.A. and Kastin, A.J. (1983). Aluminium increases permeability of the blood–brain barrier to labelled DSIP and $\beta$-endorphin: possible implications for senile and dialysis dementia. *Lancet*, **ii**, 1227–1229
42. Banks, W.A. and Kastin, A.J. (1985). Aluminium alters blood–brain barrier permeability to non-peptides. *Neuropharmacology*, **24**, 407–412
43. Banks, W.A., Kastin, A.J. and Fasold, M.B. (1988). Differential effect of aluminium on the blood–brain barrier transport of peptides, technetium and albumin. *J. Pharmacol. Exp. Therap.*, **244**, 579–585
44. Banks, W.A., Kastin, A.J. and Siznick, J.K. (1985). Modulation of immunoactive levels of DSIP and blood–brain barrier permeability by lighting and diurnal rhythm. *J. Neurosci. Res.*, **14**, 347–355
45. Kastin, A.J. and Dickson, J.C. (1987). Hypophysectomy increases Tyr-MIF-I like immunoreactivity in rat plasma. *Neuroendocrinology*, **45**, 177–181
46. Banks, W.A., Kastin, A.J. and Nager, B.J. (1988). Analgesia and the blood–brain barrier transport system for Tyr-MIF-I enkephalin: evidence for dissociation. *Neuropharmacology*, **27**, 175–179

47. Banks, W.A. and Kastin, A.J. (1985). Aging, peptides and the blood–brain barrier: implications and speculations. In: *Treatment Development Strategies for Alzheimers Disease*, eds. Crook, T., Bartus, R., Ferris, S. and Gerhson S., M. Powley Assocs., Madison CT, 245–265

# 10 Strategies for peptide delivery to the brain

There are various strategies for peptide delivery to the central nervous system (Table 10.1)[1,2], which include *invasive*, *pharmacological* and *physiological*-based procedures. Invasive procedures involve implantation of an intraventricular catheter, followed by pharmaceutical infusion into the ventricular compartment, similar to that reported for the intracranial drug infusion in patients with Alzheimer disease[3]. Due to deep convexities of the human brain surface, the distances of most brain regions from the ependymal and/or cortical surface of the CSF compartment are too large for the diffusional movement of molecules. It has been estimated, based on diffusion coefficients of $2 \times 10^{-6}$ cm $s^{-1}$ for a peptide with an average molecular weight of 2000–5000, that the distance moved by free diffusion during 1 h would be approximately 1.2 mm. Taking into account that distances of brain regions from the site of injection may be within the range of 0.5–1.0 cm, as well as the possible degradation rate of the peptide which may be rapid ($t < 10$ min), then it becomes clear that this route for peptide delivery to the brain may be less than adequate. Thus intraventricular administration of drugs in humans by implantable pump may be useful for diseases that have predilection for the meninges, e.g. leukemic infiltration, but are not useful for delivery of drugs to the brain parenchyma, due to high efflux from the ventricular compartment into the superior sagittal sinus, which is faster than diffusion of drug into brain parenchyma[2]. One variation of the intrathecal approach may be to administer the peptide via nasal spray, since intranasally administered substances are directly taken up by the brain as a consequence of the continuity of the subarachnoid space of the olfactory lobe and the submucous space of the nose.

Another invasive approach would be administration of the peptide in hyperosmotic carotid injections, containing osmotically active substances such as mannitol and/or arabinose, which result in shrinkage of the brain endothelial cells and transient opening of the barrier[4]. This approach has been applied in chemotherapy of brain tumors, but the size of pores generated by hyperosmotic opening is not large enough to allow passage of

**Table 10.1** Possible strategies for peptide delivery to brain (according to Pardridge (1986, 1988)[1,2])

| *Group* | *Strategy* |
|---|---|
| Invasive | Intraventricular infusion<br>Hyperosmotic carotid infusions |
| Pharmacological | Liposomes<br>Latentation (lipid-soluble prodrugs) |
| Physiological | Chimeric peptides |

enzymes such as decodaminidase A through the blood–brain barrier. Also, the toxicity of such a procedure is high and the incidence of seizures may be as great as 20%.

Among the pharmacological strategies, the use of liposomes for peptide delivery across the blood–brain barrier has been shown to be inadequate, since liposomes, although avidly taken up by the cells of the reticuloendothelial system of liver and spleen, are not measurably transported through the blood–brain barrier[5]. A potentially more powerful approach is *latentiation*, which involves conversion of the functional group of the peptide molecule from water-soluble into the lipid-soluble. One form of latentiation with dipeptides is the formation of diketopiparzines. Coupling of the terminal carboxy and amino groups results in a large increase in the lipid solubility of these compounds, due to the loss of several functional hydrogen bonds of the parent molecule[6]. Thus, for example, non-cyclised peptides, such as glycyl-leucine and glycyl-phenylalanine are poorly transported across the blood–brain barrier[7] while the cyclised peptide or diketopiperazine of the oxytocin C-terminal leucyl-glycine[8], or C-terminal histidyl-proline of TRH[9] were transported readily across the blood–brain barrier. The major factors limiting the free diffusion of peptides across the cerebral endothelial cells are amide linkages which easily form hydrogen bonds and these decrease the permeability of the compounds by a log order for each pair of hydrogen bonds on the molecule[2]. However, high lipid solubility is not always correlated with the high blood–brain barrier permeability to the peptide. Thus, for example, a cyclic endecapeptide of fungal origin, *cyclosporin*, has an extremely high octanol/Ringer partition coefficient, approaching that of the steroid hormones, but is very poorly transported across the blood–brain barrier[11]. This example illustrates that the nature of interaction with the blood–brain barrier is critical and it may change the lipid solubility of a compound to such an extent that the transport and permeability are comparable to that of an inert polar molecule. Also, formation of lipid-soluble technetium derivatives from the water-soluble ones makes possible an application of neuro-imaging procedures to measure regional changes in cerebral blood flow which may parallel regional differences in brain function. The placement of three to four

**Table 10.2** Chimeric peptides (according to Pardridge (1985, 1988)[2,13])

| *Transportable peptide* | *Non-transportable peptide* |
|---|---|
| Insulin | Diphtheria toxin, fragment A<br>Ricin, B chain<br>$\alpha_2$-haemoglobulin<br>Cholera toxin, component B |
| Transferrin | Peroxidase<br>Diptheria toxin<br>Ricin, A chain |
| Cationized albumin | $\beta$-endorphin |

carbon primary or secondary amine tail on the lipid-soluble technetium agent may enhance the sequestration of the molecule in the brain, similar to that found for propranolol or lidokaine[12].

Formation of *chimeric peptides*[13] represents an example of physiologically based strategy, employing the possibility of some peptides that are measurably transported by receptor-mediated transcytosis to conjugate to the others which are poorly transported. The chimeric peptide is formed when a transportable peptide is covalently coupled to a non-transportable peptide, preferably using a cross-linking reagent that can be cleaved once in brain tissue. Some examples are given in Table 10.2. This, chimeric peptide may be transported using the receptor-mediated transfer of naturally transported ones, such as insulin, transferrin or cationized albumin. For example, $\beta$-E-(N-succinidyl 3-(2-pyridylthio))propionate-cationized albumin chimeric peptide is rapidly taken up and endocytosed by isolated bovine brain capillaries, via a process that is saturated by unlabelled cationized albumin, but not by unconjugated $\beta$-E and/or native albumin[14].

## REFERENCES


1. Pardridge, W.M. (1986). Receptor-mediated transport through the blood–brain barrier. *Endocrin. Res.*, **7**, 314–330
2. Pardridge, V.M. (1988). Recent advances in blood–brain barrier transport. *Ann. Neurol. Pharmacol. Toxicol.*, **28**, 25–39
3. Harbaugh, R.E., Roberts, D.W., Coombs, D.W., Saunders, R.L. and Reeder, T.M. (1984). Preliminary report: intracranial cholinergic drug infusion in patients with Alzheimer's disease. *Neurosurgery*, **15**, 514–517
4. Neuwelt, E.A. and Rapoport, S.I. (1984). Modification of the blood–brain barrier in the chemotherapy of malignant brain tumors. *Fed. Proc.*, **43**, 214–221
5. Patel, H.M. (1984). Liposomes: bags of challenge. *Biochem. Soc. Trans.*, **12**, 333
6. Pardridge, W.M. (1986). Chemistry of cyclosporin. In: *Nutrition and the Brain*, eds. Wurtman, R.J. and Wurtman, J.J. Raven Press, New York, **vol.7**, 199–203
7. Zlokovic, B.V., Begley, D.J. and Chain, D.C. (1983). Blood–brain barrier permeability to dipeptides and their constituent amino acids. *Brain Res.*, **271**, 65–71

8. Hoffman, P.L., Walter, R. and Bulat, M. (1977). An enzymatically stable peptide with activity in the central nervous system: its penetration through the blood–CSF barrier. *Brain Res.* **122**, 87
9. Peterson, J.S., Kalivas, P.W. and Prasad, C. (1984). Cyclo (His-Pro) (cHP) regulates striatal dopaminergic function. *Soc. Neurosci. Abstr.*, **10**, 1123
10. Pardridge, W.M. and Mietus, L.J. (1979). Transport of steroid hormones through the rat blood–brain barrier. Primary role of albumin-bound hormone. *J. Clin. Invest.*, **64**, 145
11. Cefalu, W.T. and Pardridge, W.M. (1985). Restrictive transport of a lipid-soluble peptide (cyclosporin) through the blood–brain barrier. *J. Neurochem.*, **45**, 1954
12. Pardridge, W.M., Sakiyama, R. and Fierer, G. (1984). Blood–brain barrier transport and brain sequestration of propanolol and lidocaine. *Am. J. Physiol.*, **247**, 582–588
13. Pardridge, W.M. (1985). Strategies for drug delivery through the blood–brain barrier. In: *Directed Drug Delivery: A Multidisciplinary Problem*, eds. Borchardt, R.T., Repta, A.J. and Stella V.J. Humana Press Inc., Clifton, NJ, 83
14. Kumagai, A.K., Eisenberg, J. and Pardridge, W.M. (1987). Absorption-mediated endocytosis of cationized albumin and a beta-endorphin-cationized albumin chimeric peptide by isolated brain capillaries. Model system of blood–brain barrier transport. *J. Biol. Chem.*, **262**, 15214–19

# Index


'A' system of amino acid transport, 38, 39, 41–2
A4 amyloid peptide, 169
acetylcholine, somatostatin and, interactions, 65–6
acquired immune deficiency syndrome, peptide-T and, 134
adenohypophysis
  peptides from, in CSF, 140
  prolactin release from, 69–70
adenylate cyclase
  in cerebral microvessels, 70, 157
  in protein transport regulation, 157
adrenal cortex in stress, 69
adrenocorticotrophin (ACTH), *see also* melanocyte stimulating hormone/ACTH4–10
  affective disorders and, 167
  biosynthesis, 114
  effects, 51
  release, 167, 181
    stress-induced, 69
  synthetic analogue, effects, 58
adrenocorticotrophin-related peptides in dementia, effects, 169–70
affective disorders, peptides and, 165–8
aggression, peptide administration affecting, 55
AIDS, peptide-T and, 134
D-Ala-peptide-T-amide, kinetics/transport, 134
albumin
  cationized, in a chimeric peptide, 191
  kinetics/transport, 152, 156
    in multiple sclerosis rat model, 184
  as a plasma-carrier protein, 153
alcohol intake, peptide administration affecting, 55
alkaloids, opiate, endogenous mammalian, 98
aluminium neurotoxicity, possible mechanism, 184
Alzheimer's disease, peptides and, 169
amino acid(s), exchange/transport, 31–42
  locations, 31
amino acid carriers, 32–3, 36, *see also specific carriers*
  types, 32–3
aminopeptidases, 141, 142
AMP, cyclic (cAMP)
  in endothelial cells, levels, 70
  formation, vasoactive intestinal peptide and, 65
  in protein transport regulation, 157
amphetamine model of experimental psychosis, 183
amygdala, corticotrophin releasing factor binding sites in, 102
amyloid peptide, A4, 169
anaesthesia effects on amino acid transport, 35
angiotensin I, 108–9
angiotensin II, 108–10
  anatomical distribution, 108–10
  biosynthesis/processing system, 114
  effects, 51, 55, 73, 74
  receptors, 73
  synthetic analogues, effects, 57, 58
angiotensin III, 108, 109
anionic amino acid transport systems, 33
antidiuretic hormone, *see* vasopressin
apolipoprotein E, 140
aqueous humour, Davson's kinetic analysis of solute entry into, 18–21
arachnoid, 7
  CSF and, amino acid exchanges between, 42
arginine uptake by blood–brain barrier, kinetic constants, 37
ASC system of amino acid transport, 39
astrocytes, amino acid transport measurements employing, 39
astrocytic end feet, 3
autocrine functions, 92
autonomic ganglia, peptide and monoamine co-localization in nerve ending storage sites in, 64
autoradiographic methods of peptide detection, 94, 107

basal ganglia

corticotrophin releasing factor binding sites in, 102
neuropeptides in, 170, 171
behavioural effects of peptides, 50, 55
blood
to brain, molecule movement from, methods of studying, 14
brain to, peptide transport, 134–6
CSF to, molecule movement from, methods of studying, 15, 137–8
blood–brain barrier
cellular interfaces of, peptide interactions with, 123–42
definition, 1
functions, methods for studying, 13–27
*in vitro*, 13, 15–18
*in vivo*, 13–14, 14–15, 18–27
mechanisms, 6
transport across, *see* transport
blood–CSF barrier
blood–tissue interface of, interactions at, 137–8
immunoglobulins and the, 181–2
in multiple sclerosis rat model, 184
blood pressure, peptide effects on, 56
bombesin/bombesin-like peptides, 111
anatomical distribution, 111
effects, 51, 55
receptors, 111
bradykinin
effects, 70, 73
synthetic analogues, effects, 57, 58
brain, *see also* fetal cerebrospinal fluid–brain barrier
areas/regions
lacking a blood–brain barrier, 1
special, 8
blood to, molecule movement from, methods of studying, 14
blood to, peptide transport, 134–6
Davson's kinetic analysis of solute entry into, 18–21
disease/disorders, peptides and, 165–73, 183–4
experimental models, 182–4
intact protein entry into, structural aspects, 154–6
perfusion, *see* perfusion
sampling techniques after intravenous injection of test solute, 14, 23–5
vascular perfusion method, 14, 25–7
brain uptake index (BUI), 20, 32, 34

calcitonin, biosynthesis, 113–14
calcitonin gene–related peptide
biosynthesis, 113–14
effects, 51, 55
calcium
mobilization, 70
storage, 72
calcium-sensitive guanylate cyclase, in protein transport regulation, 157
capillary
cerebral
brain side of, amino acid transport and the, 38–42
continuous, 1–3
endothelium, *see* endothelium
fenestrated, 3–4
isolated, amino acid transport measurements employing, 38
permeability characteristics, 4
systemic, 1
carboxypeptidases in peptide processing, 114
cardiovascular effects of peptides, 56
carotid body, peptide and monoamine colocalization in nerve ending storage sites in, 64
carotid injections of peptides
hyperosmotic, 189–90
Olendorf's, 14, 19–21
carriers, amino acid, *see* amino acid carriers
cationic amino acid transport systems, 33
caudate nucleus, enkephalin neurons in, 96
cell membrane, electrophysiological effects of peptides on, 62, 63
central nervous system, peptides in
anatomical distribution, 94–113
effects, 47–60
cerebral capillary, *see* capillary
cerebral microvessels, *see* microvessels
cerebrospinal fluid (CSF), 7–8, *see also* blood–CSF barrier
amino acid transport relating to the, 40–2
arachnoid and, relations between, 42
to blood, molecule movement from, methods of studying, 15
Davson's kinetic analysis of solute entry into, 18–21
in development, source and type, 158–60
interstitial fluid (of brain) and, relations between, 40
peptides
in affective disorders, 165–6
plasma peptides and, relationship between, 138–40
proteins in, 157–60

secretion, peptide effects on, 75
cerebrospinal fluid–brain barrier, fetal, 160
cerebrovascular permeability to non-electrolytes, measurement of lower limits, 24
cervical ganglion, superior, peptide and monoamine co-localization in nerve ending storage sites in, 64
chimeric peptides, 191
cholecystokinin/cholecystokinin-related peptides, 112
  anatomical distribution, 112
  antipsychotic action, 168
  receptors, 112
choroid plexus(es), 7–8
  amino acid transport involving the, 40–2
  isolated, 15
  peptide transport involving the, 137, 138
circle of Willis, 35
circumventricular organs, 8
  peptide receptors at, 72–3
claustrum, neuropeptides in, 170
codeine, endogenous mammalian, 98
corpus striata, neuropeptides in, 170
corticosterone, kinetics/transport, 153
corticotrophin releasing factor, 100–1
  affective disorders and, 167
  anatomical distribution, 100–1
  effects, 51, 69, 167, 181
  receptors/binding sites, 100–1, 102
Crone's equation, 21, 23
Crone's single pass method, 32, 34
CSF *see* cerebrospinal fluid
cultures of endothelial cells, *see* endothelial cells
cyclic nucleotides in protein transport regulation, 156–7, *see also* AMP, cyclic
cycloleucine uptake by blood–brain barrier, kinetic constants, 37
cyclosporin, kinetics/transport, 133
cytochalasin B, glucose carrier studies using, 127

Davson's kinetic analysis of solute entry
  into aqueous humour, 18–21
  into brain, 14, 18–21
  into CSF, 18–21
degradation of peptides, enzymatic, 141–2
dehydration, angiotensin II receptors and, 110
delivery strategies for peptides, 189–91
delta sleep inducing peptide, *see* sleep inducing peptide, delta
dementia, peptide effects in, 169–70
depolarization-induced release of peptides, 62
depression, peptides and, 165–8
development, proteins in CSF during, source and type, 158–60
diabetes, experimental models, insulin binding in brain microvessels in, 183–4
diacylglycerol, 66, 67
diffusion in amino acid transport, 36
diketopiperazines, 142, 190
dipolar amino acids, transport systems, 33
disease/disorders, brain-related, peptides and, 165–73
dopamine, actions, 67–8
dopaminergic pathway interactions
  with neurotensin, 64–5
  with peptidergic pathways, in extrapyramidal disorders, 172
drinking, peptide administration affecting, 55
drug delivery strategies, 189–91
DSIP *see* sleep inducing peptide, delta
dura, 7
dynorphin
  effects, 51
  synthetic analogue, effects, 57, 58

electrophysiological effects of peptides, 62, 63
eledoisin, effects, 51
encephalomyelitis, experimental allergic, 184
endocrine functions, 92
endocytosis, receptor-mediated, 150, 154
endopeptidase 24.11, 141
endorphin(s)
  effects, 51, 52, 55
  synthetic analogue
    effects, 57, 58
    kinetics/transport, 129
β-endorphin
  affective disorders and, 167–8
  biosynthesis, 114
  effects, 51, 55, 68–70, 167–8
endothelial cells, capillary/microvessel
  gaps between, 3
  monolayer cultures, 17
    amino acid transport measurements employing, 38–9
  polyphosphoinositide metabolism in, 70
endothelium, cerebral capillary
  genetic expression in, 76
  morphology, 1–4
  peptide effects on, 73, 76

permeation routes through, 4–6
enkephalin(s), 95–9
anatomical distribution, 95–9
biosynthesis/processing system, 114
effects, 52, 74
enzymatic degradation, 141
enkephalin (Leu)
anatomical distribution, 97, 98
kinetics/transport, 131, 134–5, 137, 183
synthetic analogues
actions/effects, 58–9, 60, 172
binding sites, 98
kinetics/transport, 137
enkephalin (Met)
anatomical distribution, 97, 98
synthetic analogues
effects, 59, 60
kinetics/transport, 129
enkephalinergic pathways in CNS, 95, 96–7, 98
enzymes
as barrier mechanisms, 6
in cerebral microvessel preparations, demonstration/studies, 16
peptide-degrading, 141–2
peptide-processing, 114
ependyma, 8

feeding behaviour
cholecystokinin and, 112
peptide administration affecting, 55
fetal cerebrospinal fluid–brain barrier, 160
fibrillar matrix, 5
FK-33-824 effects on choreic movements, 172
frontoparietal cortex, corticotrophin releasing factor binding sites in, 102

gaps between endothelial cells, 3
gastric secretion, peptide effects on, 56–7
gastrin, effects, 52
gastrin releasing peptide, effects, 52, 55
gastrointestinal hormones, 92, 93
gene site, nerve cell function/activity modulated at, 63
glia
amino acid exchange between interstitial fluid, neurons and, 39
projections of specialized, astrocytic end feet as, 3
glioma cells, amino acid transport measurements employing, 39
globulins in CSF, 158
globus pallidus, enkephalins in, 96
glucagon, effects, 52
glucose transporter/carrier, 18, 127
gonadotropin releasing hormone, 99
growth factors, 93
growth hormone, regulation, 67, 67–9
growth hormone releasing factor
effects, 52
release, 68
GTP binding coupling protein complex, 66
guanylate cyclase in protein transport regulation, 157

heart rate, peptide effects on, 56
higher integrative brain functions, peptides and, 50–6
histamine in protein transport regulation, 157
histidine uptake by blood–brain barrier, kinetic constants, 37
histidyl-proline diketopiperazin, 142, 190
hormonal changes, stress-induced, opiate role in, 69
horseradish peroxidase, kinetics/transport, 155–6, 159–60
humoral pathway of neuro-immunomodulation, 179
Huntington's disease, peptides and, 171–3
5-hydroxytryptamine (serotonin), vasoactive intestinal peptide interactions with, 65
hyperosmotic carotid injections of peptides, 189–90
hypertonicity, angiotensin II receptors and, 110
hypophyseal peptides, 93
hypothalamo-hypophyseo-adrenal axis, stress response and the, 69
hypothalamo–hypophysiotropic axis, 91
hypothalamo–hypophysiotropic hormones, 91, 93
hypothalamus
corticotrophin releasing factor binding sites in, 102
enkephalins cells in, 96–7
peptides interactions with, 66–70

immune system, 179–82
immunocytochemical methods of peptide detection, 94–5
immunoglobulin(s)
and the blood–CSF barrier, 181–2
in CSF, 158
immunoglobulin G, kinetics/transport, 152, 182
*in vitro* methods for studying blood–brain barrier function, 13, 15–18

*in vivo* methods for studying blood–brain barrier function, 13–14, 14–15, 18–27
indicator diffusion technique, 21–3
injection methods (for peptides etc.)
  experimental, 14, 15, 19–21, 23–5
  in humans, 189–90
inositol triphosphate formation, 70
insulin, 107–8
  anatomical distribution, 107–8
  biosynthesis, 114
  in a chimeric peptide, 191
  kinetics/transport, 150–1
  receptors/binding sites, 107–8, 150–1
    in experimental models of diabetes, 183–4
insulin-like growth factors
  anatomical distribution, 108
  kinetics/transport, 151–2
insulin-like growth factor-I, 151–2
insulin-like growth factor-II, 151–2
integrative brain functions, higher, peptides and, 50–6
interstitial fluid (brain)
  amino acid transport relating to, 39, 40
  CSF and, relations between, 40
intracarotid injection technique, Olendorf's, 14, 19–21
intracerebroventricular injection techniques, 15, 189
intravenous infusion technique of measuring amino acid uptake, quasi-steady state, 34
intravenous injection of test solute, brain sampling techniques after, 14, 23–5
intraventricular injection methods, 15, 189
invasive methods of peptide delivery to humans, 189–90
isoleucine uptake by blood–brain barrier, kinetic constants, 37

$K_D$ for amino acids, determination, 35–6, 37
$K_{in}$, determination, 24–5, 25–6, 131
$K_m$ for amino acids, determination, 35–6, 37
kinetic *in vivo* studies of solute transport across blood–brain barrier, long–term, 128–34

L system of amino acid transport, 38, 39
lamina terminalis
  angiotensin II receptors in, 110
  organum vascularis of, *see* organum vascularis of lamina terminalis
large neutral amino acid carriers, 41
latentiation, 190–1
learning, vasopressin effects on, 103
leucine, blood–brain barrier permeability and the effects of, 76
leucine uptake by blood–brain barrier, kinetic constants, 37
light, blood–brain barrier and the effects of, 184–5
lipid-mediated transport processes, 152
lipid-soluble prodrugs, 190–1
liposomes for peptide delivery, 190
lipotropic pituitary hormone, biosynthesis, 114
litorin, effects, 52, 55
local routes of peptide administration for studying actions, 49
luteinizing hormone, regulation, 67–9
luteinizing hormone releasing hormone, 99–100
  anatomical distribution, 99–100
  effects, 52, 67, 68
lymphokines, 180
lysine uptake by blood-brain barrier, kinetic constants, 37

macromolecular transport, 149–60
magnesium activated phosphodiesterase in protein transport regulation, 157
mania, peptides and, 165, 166
matrix, fibrillar, 5
medulla oblongata
  peptide and monoamine co-localization in nerve ending storage sites in, 64
  thyrotrophin releasing hormone and substance P-like peptides in, 99
melanocyte stimulating hormone (MSH/$\alpha$-MSH)
  biosynthesis, 114
  effects, 52
    in Parkinson's disease, 172
  synthetic analogue, effects, 57, 59
melanocyte stimulating hormone/ACTH4–10
  effects, 52
    in depression, 168
  synthetic analogue, effects, 57, 76, 170
melanotrophin release-inhibiting factor-1
  effects, 52, 55
    in Parkinson's disease, 172
  kinetics/transport, 135–6
  synthetic analogues, kinetics/transport, 135–6, 184–5
membrane, cell, electrophysiological effects of peptides on, 62, 63
memory, vasopressin and, 50
meninges, components, 7

mesolimbic dopaminergic pathway interactions with neurotensin, 64–5
methionine uptake by blood–brain barrier, kinetic constants, 37
N-methyl-4-phenyl-1,2,3,6-tetrahydropyridine-induced syndrome resembling parkinsonism, 183
microvessels, cerebral
  endothelial cell cultures from, 17
  in experimental models of diabetes, 183–4
  glucose transporter, 18
  isolated, techniques employing, 15–16
  peptide and protein interactions/effects with, 70–2, 113, 150–2
monoamines, peptides and, co-localization of storage sites in nerve endings, 63, 64
morphine, endogenous mammalian, 98
multiple sclerosis rat model, blood–CSF barrier permeability to albumin in, 184
multivesicular bodies, transport in, 156

$Na^+$-dependent dipolar amino acid transport systems, 33
$Na^+$-independent dipolar amino acid transport systems, 33
neural pathway of neuro-immunomodulation, 179
neurocrine functions, 92
neuroendocrine axis, peptides and the, 66–70
neurohypophyseal hormones, 92, 93
neuromodulators, peptides as, 60–6
neuron(s), amino acid exchange between interstitial fluid, glia and, 39
neuronal perikarya, delta sleep inducing peptide in, 106
neuropeptide(s), *see* peptides
neuropeptide Y, effects, 52, 55
neurotensin
  effects, 52, 55, 64–5
  nigrostriatal and mesolimbic dopaminergic pathway interactions with, 64–5
neurotransmitters
  actions in blood versus brain, 6
  peptides and, interactions in Huntington's and Parkinson's disease, 171–2
  peptides as, 60–6
neutral amino acid carriers, large, 41
nigrostriatal dopaminergic pathway interactions with neurotensin, 64–5

nociception, bombesin-like peptides and, 111
noradrenaline, vasoactive intestinal peptide interactions with, 65
nucleus tractus solitaris, angiotensin II binding in, 109
nutrient transport systems, 126–7

Oldendorf's intracarotid injection technique, 14, 19–21
olfactory system, vasoactive intestinal peptide and the, 113
opiate(s)
  receptors, 95, 97, 98
  stress-induced hormonal changes and the role of, 69
opiate alkaloids, endogenous mammalian, 98
opioid (peptides), 93, 95–9
  anatomical distribution, 95–9
  binding sites, 97
  kinetics/transport, 128–31, 134–5
  physiological role, 98
  synthetic analogues, 58–9, 129–33
δ-opioid agonist, effects, 74
ORG2766, effects, 57, 76
  in dementia, 170
organum vascularis of the lamina terminalis
  angiotensin II binding in, 109
  luteinizing hormone releasing hormone in, 100
ornithine uptake by blood–brain barrier, kinetic constants, 37
orosomucoid as a plasma-carrier protein, 153
oxytocin, 105
  anatomical distribution, 105
  binding sites/receptors, 105
  effects, 52, 105

pain perception, enkephalins and other peptides involved in, 96
paired-tracer indicator dilution method, 14, 23
pallidum, ventral, neuropeptides in, 170
pancreatic polypeptide, effects, 52, 55
paracrine functions, 92
parathyroid hormone, effects, 70
paraventricular nuclei
  angiotensin II binding in, 109, 110
  enkephalins cells in, 96–7
Parkinson's disease, peptides and, 171–3
peptidases, 141–2
peptidergic pathways, 91–115

in extrapyramidal disorders, dopaminergic pathway interactions with, 172
recognition, 61–2
peptide(s) (neuro-/brain), 47–77, 91–115, 165–73, 189–91
abbreviations, 48
actions/role, 47–77
routes of administration for studying, 48–50
versus actions in blood, 6
anatomical distribution in CNS, 94
assay first – factor later, 61
biosynthesis, 113–14
brain disorders and, 165–73, 183–4
brain-derived, 115
carriers, 154
chimeric, 191
circulating (synthesized outside), extra-brain pool of, 114–15
classification/categorization, 93
delivery strategies, 189–91
enzymatic degradation, 141–2
factor first – function later, 61
monoamines and, co-localization of storage sites in nerve endings, 63, 64
same, in different tissues, primary structure, 94
synthetic analogues, 57–60
transport, *see* transport
peptide-T, 134
perfusion
brain
amino acid uptake determinations via, 34
vascular, method of, 14, 25–7, 130–4
from ventricles to aqueduct of Sylvius, 15
ventriculo-cisternal, 15, 40, 134–6
perikarya, neuronal, delta sleep inducing peptide in, 106
peripheral actions of peptides, 56–7
peripheral routes of peptide administration for studying actions, 49
permeability, blood–brain barrier, 4
alterations, 75–7
to non-electrolytes, measurement of lower limits, 24
to peptides, 75–7, 126, 131
measurement, 126
permeation routes through cerebral capillary endothelium, 4–6
pharmacological methods of peptide delivery, 190–1
phenylalanine uptake by blood–brain barrier, kinetic constants, 37
phosphodiesterase in protein transport regulation, 157
phospholipase A, 66
phospholipase $A_2$, 66, 67
phospholipase C, 66, 67
phosphoproteins, 50–55 kDa, 17, 77
physiological methods of peptide delivery, 191
pia, 7
pineal gland hormone, 92
pituitary hormone secretion, peptides and the, 66–70
plasma-carrier proteins, 153
plasma peptides, CSF peptides and, relationship between, 138–40
polyphosphoinositide metabolism in endothelial cells, 70
positron emission tomography, blood–brain barrier function studies using, 14
potassium channel opening and closing, electrophysiological effects of peptides coupled to, 62
preoptic nucleus, median, angiotensin II binding in, 109, 110
processing of peptides and their mRNAs, 113–14
prodrugs, lipid-soluble, 190–1
prolactin
plasma and CSF levels, relationship between, 139
regulation, 69–70
pro-opiomelanocortin peptides, processing, 114
protein(s)
in CSF, *see* cerebrospinal fluid
transport of, 149–60
regulation, 156–7
protein(s), transporter (blood–brain barrier proteins), 8–9, 128
14–18 kDa, 17–18, 77
46 kDa, 18, 77
50–55 kDa phospho-, 17, 77
isolation, 17–18
protein-bound molecules, kinetics/transport, 152–3
psychoendocrinology, 165
psychosis, experimental, amphetamine model of, 183

quantitative autoradiographic methods of peptide detection, 94, 107

radioimmunoassays, peptide, 94, 95
receptor(s), *see also specific receptors*

endocytosis mediated by, 150, 154
nerve cell function/activity modulated by, 63
peptide
in cerebral microvessels, 70–2
at circumventricular organs, 72–3
peptide transporters and, at blood–brain barrier, 126
transcytosis mediated by, 154
recombinant DNA technology in detection of brain-specific mRNA, 61, 113–14
renin, 109
renin–angiotensin system, 108
respiration, peptide effects on, 57
Ringer and CSF, amino acid flux between, 41
mRNA, brain-specific, recombinant DNA technology in detection of, 61, 113–14

septal areas, corticotrophin releasing factor binding sites in, 102
serotonin, vasoactive intestinal peptide interactions with, 65
single pass method, Crone's, 32, 34
sleep, peptide administration affecting, 55
sleep inducing peptide, delta, 106
anatomical distribution, 106
effects, 51, 75
function, 106
kinetics/transport, 132–3, 136, 137, 139
plasma and CSF levels, relationship between, 139
sodium, *see* $Na^+$
solute
entry, kinetic analysis of, *see* Davson's kinetic analysis of solute entry
test, intravenous injection, brain sampling techniques after, 14, 23–5
somatostatin (SRIF), 101–2
acetylcholine and, interactions, 65–6
anatomical distribution, 101–2
effects, 52–3, 55, 68
in Huntington's disease, 171
somatotropin release inhibiting factor (SRIF), *see* somatostatin
stereospecificity in amino acid transport, 36
storage sites for peptides and monoamines in nerve endings, co-localization, 63, 64
stress, hormonal changes induced by, opiate role in, 69
striate cortex, corticotrophin releasing factor binding sites in, 102
subfornical organ, angiotensin II binding in, 109, 110
substance P, 96
anatomical distribution, 96
biosynthesis, 114
effects, 52
Parkinson's disease, 173
sugar uptake by choroid plexus out of CSF, 40–1
supraoptic nuclei
angiotensin II binding in, 109, 110
enkephalins cells in, 96–7
Sylvius, aqueduct of, perfusion from ventricles to, 15
sympathetic ganglia, peptide and monoamine co-localization in nerve ending storage sites in, 64
systemic routes of peptide administration for studying actions, 49

tanacytes, trancytosis through, 140
technetium derivatives, lipid-soluble, 190–1
tegmental area, ventral, peptide and monoamine co-localization in nerve ending storage sites in, 64
temperature, body, peptide effects on, 57
testosterone, kinetics/transport, 153
thalamus, corticotrophin releasing factor binding sites in, 102
threonine uptake by blood–brain barrier, kinetic constants, 37
thyroid hormones, transport systems, 126–7
thyrotrophin releasing hormone
affective disorders and, 167
enzymatic degradation, 141
in Huntington's disease, 171
identification and localization, 91, 99
kinetics/transport, 130, 132, 133, 137–8
transcytosis, 154, 156
receptor-mediated, 154
tanacyte-mediated, 140
transferrin
in a chimeric peptide, 191
kinetics/transport, 152, 156
transport (of molecules across barriers), 13–27, 123–40
methods of studying, 13–27
modulating influences, 184–5
proteins/large peptides, 149–60
small peptides, 123–40
systems (transporters), 126–8
transport proteins, *see* proteins, transporter
tryptophan uptake by blood–brain barrier, kinetic constants, 37

tyrosine uptake by blood–brain barrier, kinetic constants, 37

unidirectional transfer constant ($K_{in}$), determination, 24–5, 25–6, 131
Ussing chambers, 17

$V_{max}$ for amino acids, determination, 35–6, 37
valine uptake by blood–brain barrier, kinetic constants, 37
vascular brain perfusion method, 14, 25–7, 130–4
vasoactive intestinal peptide, 112–13
  affective disorders and, 165–6
  anatomical distribution, 112–13
  effects, 54, 70, 74, 75
  neurotransmitter interactions with, 65
  receptors, 112–13
vasoconstriction, peptide-mediated, 73–4
vasomotor effects of peptides
  cerebral, 73–5
  CSF secretion and, 75
vasopressin (antidiuretic hormone), 103–4
  anatomical distribution, 103–4
  binding sites/receptors, 71–2, 73, 103–4, 105
  in dementia, 169, 170
  effects of, 54, 55, 70, 75, 169
    memory and learning and the, 50, 103
  kinetics/transport, 136
    vasopressin arginine, 103–4
  affective disorders and, 167
  agonists/antagonists, 103, 104, 105
  anatomical distribution, 103–4
  binding/receptors, 71–2, 103–4
  effects, 54, 71–2, 73, 74, 75–6
  kinetics/transport, 133, 136, 183
  plasma and CSF levels, relationship between, 139
  synthetic analogues
    binding sites/receptors, 104
    effects, 57, 59
vasopressin lysine
  affective disorders and, 167
  effects, 54, 55, 73, 74
  kinetics/transport, 136
vasopressin-ergic fibres in brain capillaries, 71
vasotocin arginine, effects, 54, 55
vegetative functions of CNS, peptide effects, 56–7
ventral pallidum, neuropeptides in, 170
ventricle(s)
  injection methods, 15, 189
  perfusion from, to aqueduct of Sylvius, 15
ventriculo-cisternal perfusion, 15, 40, 134–6
vesicles in cytoplasm of capillary endothelial cells, 5–6
visual system, vasoactive intestinal peptide and the, 113

Willis, circle of, 35

MIX
Papier aus verantwortungsvollen Quellen
Paper from responsible sources
FSC® C105338

If you have any concerns about our products,
you can contact us on
**ProductSafety@springernature.com**

In case Publisher is established outside the EU,
the EU authorized representative is:
**Springer Nature Customer Service Center GmbH**
**Europaplatz 3, 69115 Heidelberg, Germany**

Printed by Libri Plureos GmbH
in Hamburg, Germany